69

Brings the latest space results down to Earth—a story well told.

—*John Mather, Ph.D., COBE satellite chief scientist*

Heeren uses very clever dialogues and information from interviews with world-renowned scientists to make his presentation particularly interesting and authoritative. As a person who has researched the interface between Christianity and science for 25 years, I can vouch for the scientific accuracy of the various arguments Heeren has made. I would strongly recommend this book to the scientist and the lay reader alike who are interested in learning more about scientific evidence for the existence of God.

—*Walter L. Bradley, Ph.D., professor and senior research fellow, Texas A&M University, Department of Mechanical Engineering*

To assist us in our witness, my good friend, Fred Heeren, has compiled a veritable compendium of material from careful research and reading for people like you and me. *Show Me God* is a book for our day, and all of us should have a copy to read and then relate to every strata of society. Fred wants every believer to become acquainted with the amazing scientific discoveries that can open the door to effective personal evangelism.

—*Stephen F. Olford, Th.D., Stephen Olford Center for Biblical Preaching*

In this well-researched and thoroughly enjoyable book, Fred Heeren makes contemporary cosmology accessible to the rest of us, enabling readers to see that recent discoveries in science present theologians and scientists with very encouraging prospects for dialogue.

—*Robert A. Pyne, Th.D., Dallas Theological Seminary*

The concept is unique; the task immense: to present the case for a rational faith in God by drawing both on contemporary science and on interviews with outstanding scientists, as well as on the Bible, without undue superficiality or distortion. *Show Me God* presents a wide range of ideas and evidence in an unusually readable style.

—*Walter L. Liefeld, Th.D., professor emeritus, Trinity Evangelical Divinity School*

Any seeker of truth will find much to ponder in its pages and Heeren's conclusion is both reasonable and profound.

—*J. P. Moreland, Ph.D., Talbot School of Theology, Biola University, Editor of The Creation Hypothesis*

In *Show Me God*, Heeren has faithfully brought together the two books of God [natural and special revelation] in a captivating way which provides not only a deep appreciation of the greatness of our God, but also, a very practical way to "be ready to give a defense" (1 Peter 3:15) for the hope that is in us.

—*Earl D. Radmacher, Th.D., president emeritus and distinguished professor of theology emeritus, Western Conservative Baptist Seminary*

What Are These "Wonders That Witness"?

EVERYONE HAS HEARD of the "Seven Wonders of the Ancient World," known for thousands of years as the greatest works of humankind. But even the pyramid of Cheops and the hanging gardens of Babylon lose their grandeur when compared with the timeless wonders you are about to find in this series. The seven *greatest* wonders that any human has ever contemplated are especially remarkable in light of discoveries made in this century—and in this decade. Here is Searchlight's list of the seven most impressive wonders of all time, and here is how we have incorporated them into four books:

Volume I (Cosmic Origins):
1. The wonder of the origin of the universe
2. The wonder of the universe's fine-tuning

Volume II (Examining Evolution):
3. The wonder of how life began
4. The wonder of how each life form abruptly appeared
5. The wonder of humanity

Volume III (Examining Creation):
6. The wonder of the Bible's creation account, compared with all other ancient and modern explanations

Volume IV (The Bible's Origins):
7. The wonder of the Bible's fulfilled prophecy, historical accuracy, and unique message

Though all these wonders deal with factual, undisputed events in the history of the universe, they all stand just slightly beyond what your science teacher was prepared to teach you. It turns out that science raises questions that science can't answer.

The series is called *Wonders That Witness* because all these wonders, these unique findings of science and history, bear witness to the truth of one incredibly unique book. As we will see, the book of natural revelation points to the book of special revelation, the Bible. Perhaps our world has been too quick to outgrow the advice of Francis Bacon, the father of the scientific method, who wrote:

> No one should maintain that a man can search too far, or be too well studied in the book of God's Word or in the book of God's works; divinity or philosophy; but rather let men endeavour an endless progress or proficience in both.

The scientist is possessed by the sense of universal causation. . . . His religious feeling takes the form of a rapturous amazement at the harmony of natural law, which reveals an intelligence of such superiority that, compared with it, all the systematic thinking and acting of human beings is an utterly insignificant reflection.

Albert Einstein, theoretical physicist

The laws of science, as we know them at present, contain many fundamental numbers, like the size of the electric charge of the electron and the ratio of the masses of the proton and the electron. . . . The remarkable fact is that the values of these numbers seem to have been very finely adjusted to make possible the development of life.

Stephen Hawking, theoretical physicist

A common sense interpretation of the facts suggests that a superintellect has monkeyed with physics, as well as with chemistry and biology, and that there are no blind forces worth speaking about in nature. The numbers one calculates from the facts seem to me so overwhelming as to put this conclusion almost beyond question.

Sir Fred Hoyle, astrophysicist

For the scientist who has lived by his faith in the power of reason, the story ends like a bad dream. He has scaled the mountains of ignorance; he is about to conquer the highest peak; as he pulls himself over the final rock, he is greeted by a band of theologians who have been sitting there for centuries.

Robert Jastrow, astronomer

SHOW ME GOD

SHOW ME GOD

What the Message from Space Is Telling Us About God

WONDERS THAT WITNESS
TO THE BIBLE'S TRUTH

VOLUME 1

FRED HEEREN

Forewords by
George Verwer and George Smoot

SEARCHLIGHT
PUBLICATIONS

To Patty,
an independent spirit

SHOW ME GOD
Searchlight Publications / Wheeling, IL

LIBRARY OF CONGRESS
Library of Congress Cataloging-in-Publication Data

Heeren, Fred, 1953-
    Show me God: what the message from space is telling us about God / Fred Heeren
    p.   cm.—(series title: Wonders that witness; v. 1)
    Includes bibliographical references and index.
    ISBN 1-885849-51-6
    1. Apologetics. 2. Cosmology. 3. Creation. 4. Religion and science. 5. Bible and science. I. Title. II. Series: Heeren, Fred, 1953- Wonders that witness; v. 1.
BT1102.H34 1995
261.5'5—dc20   95–3712
                                                                              CIP

Printed in the United States of America

This book is published by Searchlight Publications, a division of Searchlight, Inc., and is distributed by Day Star Productions, Inc.

For information on Searchlight's other books, study guides, dramatic recordings, tracts, church bulletin inserts, and public speaking ministries, write Searchlight, 326 S. Wille Avenue, Wheeling, IL 60090, or call 1-708-541-5200.

S OME PEOPLE find it easy to believe everything written in the Bible. Others say it is a great strain to believe a book in which almost every page contains supernatural elements. I for one find it very difficult to believe in something if I can't see it or hear it or smell it or trip over it.

I have never seen the laws of physics suspended, and I have seen enough hoaxes to doubt the claims of any who say they have. Because such miracles are completely outside of my experience (and frauds are an everyday occurrence), I find it quite a task to believe that any miracle ever happened in the past.

Yet, here we are, living proof that somehow, sometime in the past, everything must have come out of nothing—and there is just no natural way for such a thing to occur. And it turns out that physicists are saying that the universe has been very precisely "fine-tuned" to make life possible. This puts me in a bit of a dilemma. Here I am, disbelieving in miracles while the *whole universe* is apparently an indescribably enormous miracle.

How does the skeptic resolve such a dilemma? Is it resolvable? In the pages ahead, the reader will be the judge of this and many issues that, once thought through, must ultimately require every honest, thinking person to decide between two ways of living.

Hercules cluster of galaxies. —*Courtesy of Lick Observatory, University of California*

ALL I WANT is reality. Show me God. Tell me what He is really like. Help me to understand why life is the way it is, and how I can experience it more fully and with greater joy. I don't want the empty promises. I want the real thing. And I'll go wherever I find that truth system.

Lisa Baker, age 20 —quoted as one who typifies her age group, disenchanted with religion while still seeking "the answer," in George Barna's *The Invisible Generation: Baby Busters*

CONTENTS

I owe a great debt of thanks to the following scientists who granted me interviews for this book—all distinguished contributors to the scientific enterprise and many of them discoverers of this century's key findings in cosmology: **Morris Aizenman** (Executive Officer of the Division of Astronomical Science at the National Science Foundation), **Robert Gange** (the only Christian apologist I interviewed, president of the Genesis Foundation and a working scientist in thermo-physics), **Alan Guth** (father of inflationary theory in big bang cosmology), **Stephen Hawking** (widely regarded as the most brilliant theoretical physicist since Einstein, known especially for his work on singularities (both black holes and the big bang) and for his proposal of a no-boundary condition for spacetime), **Robert Jastrow** (founder of NASA's Goddard Institute, now head of the Mount Wilson Institute and its observatory), **John Mather** (NASA's COBE satellite chief scientist), **Jeremiah Ostriker** (Princeton's co-discoverer of dark matter), **Arno Penzias** (1978 winner of the Nobel prize for physics, co-discoverer of the cosmic background radiation), **George Smoot** (leader of the COBE satellite team that first detected the cosmic "seeds" that are often called "the fingerprints of God"), **James Truran** (University of Chicago's expert on internal nuclear reactions in the stars and early galaxy formation), and **Robert Wilson** (co-winner of the 1978 Nobel prize for physics for his co-discovery of the cosmic background radiation).

In my attempt to keep the story of this century's discoveries moving, I have often been forced to quote small portions from my lengthy conversations with these scientists (and small portions from their written works). Though I have made every effort to quote them in context, I should mention that the intent of their side of our conversations was certainly not to offer support for any particular metaphysical belief. I hope my clear statement of this fact will keep anyone from misunderstanding the positions taken by these contributors to 20th-century cosmology. And I hope it will further serve to show that their more supportive statements were made apart from (and sometimes in spite of) their personal beliefs, not because of them.

Lots of thought and time went into the diagrams used to explain concepts in this book—but I can't take credit for that work. Robert Jastrow was kind enough to share his photos of galaxies and their accompanying spectra from the Mount Wilson observatory. NASA's John Mather gave me the latest data from the COBE satellite. Many other diagrams are used with permission from George Smoot, originally prepared for his *Wrinkles in Time*, an excellent reference for those who would like to learn more about modern cosmology and what it is like to be a scientist.

I wish to thank the following folks who provided valuable feedback from their reading of this book and, in some cases, *using* this book in their discussion groups for skeptics: Norris Anderson, Garry Ellis, Diana Forsea, Russell Gerads, Jim Hanert, Robert Hansen, John Heeren, Bob Hillier, Dale Hugo, Dale Luetscher, Jeff Rech, and Kimo Storke.

I must also express my appreciation to John and Maurene Heeren, who provided me with the solitary lodging I needed, away from a busy home and office, during some months of writing. And special thanks must go to the selfless souls listed below who provided physical help to my wife and five children while I was off solving the riddles of the universe: Rita Donnelly, Helen Evans, Diana Forsea, Joanne Haseman, Bob Hillier, Leanne Hillier, Mary Hopkins, Kathy Horton, Anne Hugo, Donna Jacobs, Cornelia Johnson, Kathy Knottnerus, Paul Knottnerus, Adrienne Madej, Hillary Marsh, Maribeth Mazzetta, Beth Rech, Tara Seals, and Sharon White. Most of all, I thank my wife, who puts up with me.

A Christian Leader's Perspective

THIS IS A UNIQUE and powerful book that will make it extremely difficult for the reader, and especially the skeptic, to be neutral about God. And this information is not for skeptics alone. Many years ago, shortly after my conversion, I found myself overwhelmed with doubts. The Lord brought me through that crisis with a greater appreciation for the rational foundation for my faith—and to bring me through, He used a book that addressed my concerns with facts from science. I was so relieved and excited about what I read that I walked out onto the streets of Chicago and shouted to anyone who would listen, "It's true! It's true!" I know that *Show Me God* is going to help thousands of people in the same way (though they may not shout as much as I do).

If you are strong in your faith, then you need this information too. People who are truly strong in their faith don't keep it to themselves, but look for opportunities to share the Bible's good news with others. Today that means being ready to reach skeptics, many of whom believe that modern science contradicts the Bible. Why should they listen to your good news if they think it comes from a book of myths?

Fred Heeren and I argue about whether skepticism about the Bible is a greater problem in the U.S. or in Europe. I say my European skeptics are ten times tougher than his American skeptics. Whoever is right, the fact remains that Christian workers are facing a growing tide of skepticism—ironically just when discoveries in science are pointing as never before to the truth of the Bible, as this book shows.

My own testimony is that I have experienced Christ and His grace and love every day now for 40 years. I still battle doubts, but books like this, showing the consistency between faith and science, have been a phenomenal blessing and help to me. If you do not have a particular interest in science, I urge you to keep on reading this book even after you hit something you do not understand. And try to lay aside your own subjective, preconceived ideas and prejudices and be open to the clear facts set forth in these pages. You—and the skeptics in your life—will greatly benefit.

George Verwer
International Director, Operation Mobilization
London, England

A Leading Scientist's Perspective

UNTIL THE LATE 1910'S, humans were as ignorant of cosmic origins as they had ever been. Those who didn't take Genesis literally had no reason to believe there had been a beginning. This century has seen an explosion in our exploration of the Universe as new instruments and techniques have become available. Now the search for our origin is a very active and advanced field.

This cutting-edge book explores creation where science and religion ask the same questions and think the same thoughts. This is the place where all seek and see the hand of God. Everyone, layman and scientist alike, expects to find enlightenment about the big questions in the beginning of the Universe.

Fred Heeren has written an engaging and stimulating book that probes the frontier of science and faith, showing how they reconcile. In this well-written book Fred has worked to show that the Bible is in harmony with nature and that faith and science are compatible and supportive. Bible believers should have confidence that observations of the world will be consistent with the word of God. In fact *Show Me God* argues that learning more will only increase one's sense of wonder and give more confidence in one's faith.

This ground-breaking book shows that Bible believers and scientists can have a healthy and—for both—uplifting dialog, a thing I have long felt crucial for humanity. It is important for the future of society and religion to remove the barriers of misunderstanding and mistrust between faith and science. Science and religion are not to be seen as separate and even antagonistic areas. That view denies us part of our humanity just as emphasizing one part of ourselves, e.g. career, and cutting out socializing, e.g. family, friends, and community, makes us less a person. *Show Me God* is a major step bringing scientific and religious knowledge to people in a form that shows the greater completeness.

George F. Smoot
Lawrence Berkeley Laboratory
Berkeley, California

Albert Einstein, Edwin Hubble, Walter Adams, William Campbell and others outside the 100-inch telescope dome at Mount Wilson in Pasadena, California in 1931. —*Associated Press Photo*

PREFACE

A Skeptic's Questions:

Findings in science in the 20th-century have made it increasingly unreasonable to take the Bible seriously. Cosmologists can now explain the origin of the universe without God. Archaeologists have shown that the Bible is fiction. The Bible is just too primitive to make any sense in light of modern knowledge. Myth experts like Joseph Campbell have shown that the Bible can be lumped together with all the other myths that every culture invents. People used to have to look to God to explain the order in nature, but today science can explain just about everything. Whether you look at cosmology or archaeology or literary criticism, the Bible just doesn't hold up anymore. So how can you take it so seriously?

A Bible Believer's Response:

You're asking the right questions. Many people don't care whether the Bible is true or not, just so it helps meet their needs. But you're absolutely right to start with a concern about the Bible's credibility. When you begin looking a little deeper (as in the following short section), I think you'll find that in each of the areas you mentioned, the Bible has actually become *more* credible and *more* reasonable to accept as a result of knowledge gained in this century.

The actual trend in 20th-century cosmology, for example, has been to turn from a view that was inconsistent with the Bible's creation account to one that follows the Bible's scenario very well. In fact, the Bible is the only religious source coming to us from ancient times that fits the modern cosmological picture. In case after case, 20th-century archaeologists and myth experts have also been forced to turn from older views that treated the Bible as myth to ones that treat it as history. Perhaps the best way I can illustrate all this is to put these 20th-century trends in human terms. The next three pages tell the stories of three of the 20th-century's greatest thinkers: Albert Einstein, William F. Albright, and C.S. Lewis. Representing three different fields, they made discoveries that turned their thinking 180 degrees, from views that contradicted the Bible to views that lined up with what the Bible has taught all along.

Facts That Changed
Three Minds

Religion first became possible for a reasonable man
of science in the year 1927.

—Sir Arthur Eddington

Sir Arthur Stanley Eddington
(1882-1944). —*Courtesy AIP*
Emilio Segré Visual Archives

I N 1917 traditional schools of thought had convinced three great thinkers that the Bible was untrue. In that year Albert Einstein published a paper interpreting his own general theory of relativity, making it conform to the unquestioned cosmology of his day: the static universe theory. Static universe cosmology claimed that the universe is infinite in age, thus relieving the scientific community of having to deal with questions about the ultimate origin of the cosmos. According to the consensus among astronomers, stars drifted about randomly, without apparent direction toward or away from us. The nebulae were gas clouds that belonged to our own galaxy. And the Milky Way Galaxy *was* the universe. Einstein was so convinced that these views were correct that he added what is now known as a cosmological "fudge factor" to his theory in order to make it fit this favored cosmology.

In the same year, young American archaeologist William F. Albright was completing his indoctrination into German rationalistic criticism, including the Documentary Hypothesis, which taught that most of the Old Testament's "history" before the Israelite monarchy was simply legend invented about a thousand years after the supposed events. According to this approach, stories about patriarchs like Abraham, Isaac, and Jacob could not have been preserved from as early as the Bronze age. In 1918 Albright wrote an article about the mythical elements in the stories of the patriarchs, citing the Genesis 14 account of Abraham's military campaign. He concluded that the story was actually written as a political tract to rally support for the Jewish cause against the Persians, over a thousand years after the events portrayed.

And in 1917 C.S. Lewis, a young British army lieutenant fighting in World War I, turned from simple atheism to a harder-boiled atheism that would no longer brook any romantic delusions. Though his singular abilities in language, literature, and philosophy won him many honors in subsequent years at Oxford, his appreciation of all "sentimental" notions in the Greek and Latin literature he read was tempered by his materialistic worldview. Lewis's realism admitted nothing that could not be perceived by the senses. He did not yet have a doctorate to prove his training in a school of historical criticism, but he held what he believed to be the intellectual's position: that religious faith was only for uneducated, unthinking people.

During the 1920s, all three of these men had their minds changed by facts of their own discovery. In Einstein's case, the discovery had already been made back in 1915 when he had worked out his general relativity equations. And though he would not yet face the implications of his theory, other great thinkers—Eddington, Friedmann, de Sitter, and Lemaître—all found that solving Einstein's field equations demanded that the universe could not be infinitely old, but that it must have had a beginning. A universe with a beginning requires a Beginner, making it impossible to reconcile with atheism or pantheism, while pointing most naturally to the God of the Bible. The initial reaction of the scientific community was typified by Arthur Eddington's admission: "Philosophically, the notion of a beginning of the present order of Nature is repugnant to me."

Yet the facts of general relativity, combined with new observations from astronomy, made a dramatic impression even on Eddington, as demonstrated by his grandiose statement quoted at the opening of this preface. By 1927 astronomer Edwin Hubble had not only established that the nebulae contained individual stars and were themselves galaxies far outside our own galaxy, but that these galaxies were all retreating from us at high velocities. Though observable stars might be traveling in no particular direction, this turned out to be true only because they belonged to our own galaxy. Consistent redshifts in other galaxies showed them to be speeding away from us, and the most distant galaxies were retreating with the greatest velocity.

The universe was expanding and gradually decelerating, precisely as Einstein's general theory of relativity had predicted. The scientific community gradually made a 180-degree turn, from viewing the universe as infinitely old to a realization that an expanding universe required a beginning.

Only after Einstein had seen Hubble's evidence for an expanding universe (and after he had looked through Hubble's 100-inch telescope for himself) did he formally renounce his cosmological "fudge factor." Later he wrote that its addition to his equations had been "the biggest blunder of my life." After this, Einstein wrote not only of the necessity of a beginning, but of his desire "to know how God created the world. I am not interested in this or that phenomenon, in the spectrum of this or that element. I want to know His thought, the rest are details."

In 1929, the same year that Hubble stirred up the scientific community by formally publishing his observations of an expanding universe, archaeologist William F. Albright discovered and excavated a line of mounds—Bronze Age cities—forming a route for a military campaign during Abraham's time just as described in Genesis 14. The cities along this route, later called "The Way of the King," were no longer inhabited later, in the Iron Age. "Experts" had written that the entire region described had not yet been inhabited at all, and Albright admitted that he had formerly "considered this extraordinary line of march as being the best proof of the essentially legendary character of the narrative."

This was the beginning of many finds that eventually persuaded Albright, not only of the historicity of Genesis 14, but of the uselessness of the Documentary Hypothesis, a school of criticism that had been developed without benefit of archaeological support. Bronze Age inscriptions were found bearing the name "Arriyuk," the name of a participant in the military campaign of Genesis 14; and more than this, specific names of Biblical people like Abraham, Eber, Laban, and others were found in inscriptions dating from pre-2000 B.C., demonstrating their common use in that period. Archaeological investigations further showed that Genesis described cultural practices and used technical terms that had to be dated to a time long before Moses. By 1956, Albright could write, "There can be no doubt that archaeology has confirmed the substantial historicity of Old Testament tradition."

Albright's story parallels the stories of a number of other archaeologists who were trained in schools of rationalistic criticism they later destroyed with their spades. For examples, see the account of William Ramsay, trained by the Tübingen school to believe in the late composition of the New Testament (in Volume 4), or the comments of Leonard Woolley, taught to believe that the Genesis flood story had been directly derived from the Babylonian legends (in Volume 3).

It was also in 1929 that C.S. Lewis, the renowned Oxford professor, had his atheism greatly shaken while listening to an atheist friend acknowledge evidence that pointed to the historicity of the Gospel accounts. "It

During his 1929 pilgrimage to meet Edwin Hubble in California, Albert Einstein views through Mount Wilson's 100-inch Hooker telescope. Astronomers Edwin Hubble (smoking a pipe) and Walter Adams stand by.
—*Courtesy of The Huntington Library*

almost looks as if it really happened once," his friend told him. Through a combination of events that year, Lewis was "converted"—not to Christianity, but to theism. After narrowing his choices to Hinduism and Christianity, he began using his knowledge of languages and literature to make a study of the Bible, beginning with a daily reading of John's gospel in the original Greek. Lewis was amazed, later writing: "I have been reading poems, romances, vision-literature, legends, myths all my life. I know what they are like. I know that not one of them is like this."

At the same time that Einstein was peering through Hubble's telescope and formally admitting the implications of an expanding universe, Lewis was admitting the implications of Gospel documents that had all the marks of authenticity. But Lewis went a step further than Einstein, passing from theism to a personal relationship with Jesus Christ, describing the experience, not as an emotional one, but saying, "It was more like when a man, after long sleep, still lying motionless in bed, becomes aware that he is now awake."

When a mind is ready to receive them, nothing can change it like the facts. Perhaps each lifetime has its year when faith first becomes possible.

A Not-So-Cosmic Interlude

Several people have suggested that, if I really think these facts from science and history ought to be available to everyone, then this book should contain an occasional breather to give the reader's mind a rest. Never was this suggestion made more clear to me than that fateful day when I sat down at a restaurant for a breakfast meeting in Wheaton, Illinois, that great mecca of Christian organizations. At the urging of my board of directors and my publisher's advertising agency, I had arranged to meet with several marketing executives who were recognized as experts in the field of Christian publishing. Interviewing them turned out to be a very different experience from interviewing scientists, who were usually quite cautious about expressing opinions.

"There are just two things you need if you want your books to make money," said the president of one product development company. "A great title and a great cover."

I had just spent four years of my life and all my life savings on trying to get to the bottom of life's ultimate questions, and now these Christian businessmen were beginning to make me wonder if I had it all backwards.

"People don't care about life's ultimate questions," said one seasoned old marketer. "People care about money. They care about their personal appearance. They care about getting more leisure time, more physical comforts." As he continued ticking off his list, I wondered if I had foolishly squandered my time, spending my days at libraries, universities, seminaries, reading every side of my many questions, probing the minds of those who had made this century's greatest discoveries about the universe. But now, I learned, apparently there was something bigger than the universe to consider. Or at least, something more important than the book's content: the packaging. The title. The cover design.

What about content? Another executive told me he personally wasn't interested in the content. "*I* don't think about life's ultimate questions," he said, taking another bite of his poached egg. Somehow, I believed him. "I think I'm like most people," he said. "I'm a family man. I put in my time at the office, I come home. I go to church. I'm satisfied with the answers they give. Your book's got no appeal to me. No one's going to buy your books unless you appeal to some universal self-interest, some basic want. And what do people want?"

"Truth?" I ventured, just to be perverse.

"No, no—people want to dominate others. They want to emulate the admired, to *be* admired. They want more power, more popularity, more self-confidence," and he continued with another list, concluding: "You need to tell people how this will make them richer, happier, more fulfilled, how it will give them a spiritual high."

These were not words to be taken lightly. The men before me had successfully packaged many books for some of the largest religious publishers. One executive boasted that his company routinely packaged books even before they were written, relegating the content to a mere afterthought.

While plumbing the depths of the secrets of life, it seems I had naively neglected the far deeper secrets of Christian marketing. Had I but known that all my work over the years had been spent on mere incidentals, I could have been spending all that time on something truly consequential: a snappy title and cover design.

All was not lost, however. The experts hinted that, for a fee, they might be able to take my poor effort, limited as it was in popular appeal, and turn it into something that would grab people's attention. Again, they stressed, the packaging was everything. My publisher should be prepared to spend the necessary money to come up with a title and cover design that would buy the kind of expertise it took to push all the right buttons.

My present series title, they pointed out, was pathetic. *Facts That Witness to the Bible's Truth* was my working title at the time. *Facts*, they said, are boring. People don't care about facts. And many people don't know what it means to *witness* to something. For Christians who understood the word in its Biblical sense (to be Christ's witnesses), the word was a definite turn-off, since most Christians find it awkward and scary to share their faith.

I had already been concerned that my title was a bit lame, but I began to wonder what kind of title *they* would approve: *Financial Independence Through Apologetics? Achieve That Dynamic Personality Through Apologetics?*

I was grateful for their time and their advice, but I didn't hire any of them. This book has not been packaged by the experts (though I did give a lot more thought to the title, and I probably even went overboard on the subtitles). I often wonder what I missed by not having their input. In fact, lest the reader become concerned that he or she is missing what would surely have been some very interesting improvements, I have attempted to write them myself in the gray sections that are scattered throughout the text. This has not been difficult. Ever since that fateful breakfast meeting, Carl, my imaginary editor, has come to me often as a voice in my head, reprimanding me for my naiveté and making suggestions about what the book needs to reach the masses.

But the real purpose of these not-so-cosmic interludes is to bring our grand cosmic quest down to our more mundane concerns, to see how these grand questions relate to the everyday world that you and I usually have on our minds.

Carl: Yes, let's bring the subject of cosmology out of the classroom. Let's turn it into something a guy can bring up on his next dinner date.

Fred: I'm all for that.

Carl: Something you can talk to your dentist about between swishings. Something to help you make polite conversation with your mother-in-law.

Fred: Absolutely.

Carl: But first you have to get people to read this stuff. Let's face it, Fred, as your book now stands, people will be falling asleep halfway through the title. Do you know that your title has no power words in it?

Fred: It's too late to change the—

Carl: Well, for your next printing. Have you ever thought about putting the word "secrets" in the title?

Fred: Well—

Carl: The idea is to prey on the reader's feelings of inadequacy by mentioning "secrets" that he doesn't know. How about, *The Seven Secrets of a Happy Apologist?*

Fred: I don't think—

Carl: Or go for the personal prestige angle. I can see the commercial: "Amaze your friends at your next party. Astound them with your apologetics knowledge. Read *Apologetics Power*."

Fred: Perhaps we should make these non-cosmic interludes a little shorter from now on.

Carl: Cosmic! That's it! I've got the perfect title: *Cosmic Thrills!*

Fred: (sigh)

Carl: No? Oh well, if the book doesn't work out, you can still make money on the promotional tie-ins. You could have a 900 number that will give people the chance to talk to an apologist instead of an astrologist for $3.99 a minute.

Fred: I think the reader is ready to get back to something heavier again.

PART I

This Series and the Healthy Skeptic

If a man begin with certainties, he shall end in doubts;
but if he will be content to begin with doubts,
he shall end in certainties.
—Francis Bacon

Examine everything carefully;
hold fast to that which is good.
— The Apostle Paul

Why This Series?

These books have been written from the perspective of a Bible-believing skeptic.* At one time in my life I believed that the theory of evolution offered the best explanation for life on this planet and that the beginning of Genesis might be as mythical as the other ancient creation stories. I feared that science, with its big bang theory and its quantum fluctuations, had explained the origin of the universe, making God unnecessary. I assumed that most of the Old Testament stories were too remote to offer much evidence from archaeology, and worse, that there was no way of knowing whether the story of Jesus had not simply evolved over time.

Giving you the facts that changed *this* mind will require the rest of this series. Belief has never come easy for me—but the beliefs I now hold can no longer be shaken by a bad day . . . or a bad decade. I confess that I *want* to believe, but not in a fable or a lie.

---

* The rest of this section will explain why the expression "Bible-believing skeptic" is not a contradiction in terms. A skeptic's viewpoint is not a particular stretch for the author, who founded a company originally named The Association of Healthy Skeptics, Inc. Though I have made it my practice to study the Bible daily for the past twenty-six years, my heartfelt skepticism is a quality that continues to drive me—and to drive certain of my friends and family crazy. I cannot fully believe a story, whether coming from a newscaster or a friend, without first checking it for exaggerations, biases, and unsupported opinions. I cannot hear a sermon without doing the same. The Grand Skeptic of the *Wonders That Witness* audiotape set is intended to represent the voice of reason, an objective viewpoint that exercises healthy skepticism in examining each side of a question.

The Bible itself leads us to believe that, if it is actually from God, we should expect to find extra-biblical evidence for its credibility. Factual events should have left behind evidence from archaeology and written history, since "this [work of Jesus] was not done in a corner" (Acts 26:26). We should expect to find evidence from astronomy, since "the heavens declare the glory of God" (Psalm 19:1). And if the Bible is actually the Word of God, we should expect to find evidence of knowledge that cannot be explained by the surrounding cultures and times of its writers. There should be evidence of unique, supernatural *foreknowledge*, since Isaiah 46:9-10 claims: "I am God, and there is none like me. I make known the end from the beginning, from ancient times, what is still to come."

Carl: I like the way you snuck in that advertisement for the audiocassettes in that last footnote.

Fred: You're back pretty soon, aren't you?

Carl: I had to ask you: What's your marketing angle on all this skepticism business?

Fred: I just want the reader to know that the conclusions of this book aren't just based on traditions or presuppositions.

Carl: Yeah, but some people are already skeptical enough. Do you really want to encourage it?

Fred: I think many skeptics *aren't* skeptical enough. Most people who call themselves skeptics are really just skeptical about one side of an issue, because they're in love with the other side. I'm skeptical about *everything*.

Carl: Then how do you ever get to the bottom of anything?

Fred: If something's important to me, I get as close as I can to the evidence and try to see how every side lines up with it.

Carl: But what kind of evidence can you possibly have for something like the origin of the universe?

Fred: In very recent times, scientists have come up with a considerable amount of evidence that the universe is not eternal, but that it came into existence in one moment. Two of the greatest evidences for that are the fact that the galaxies are all retreating from us at tremendous speeds, and the remnant background radiation, which was predicted to look a certain way by this theory of a creation event.

Carl: Then science books serve as your final authority.

Fred: No, I'm the kind of person who has to go right to the Nobel prize-winners who discovered these things to evaluate what they really

> know about it. If the universe is supposed to be expanding, I have to see the evidence of redshifted starlight for myself. And then I have to read everything I can about any alternative explanations.
>
> *Carl:* Well, you can't blame other people if they don't want to go to the trouble of arranging interviews with Nobel laureates and studying starlight spectra for themselves.
>
> *Fred:* That's why I wrote these books.
>
> *Carl:* For the people who care about ultimate answers but are not quite as maniacal about it as you are?
>
> *Fred:* Exactly.
>
> Carl: How about this for a title? *Secrets of a Cosmic Maniac!*

The information in this series ranges from astronomy to archaeology, from biology to history. My hope is that, as much as is possible within the limits of its pages, *Wonders That Witness* will provide a handy source that can match the range of questions that come up most often concerning the credibility of the Bible and the gospel of Jesus Christ. For some, the facts presented here will help to broaden their perspective on the Bible. The arguments from these findings will serve as "equal time," showing them a side not often presented by school curricula or the media, giving more facts to help them make up their minds about whether the Bible actually is what it claims to be. For others, already convinced of the Bible's authenticity and authority, this information will be most useful as a tool to help their *friends* examine the evidence.

The ideal person to write this book would be an expert in many fields: quantum physics, big bang cosmology, relativity theory, radio astronomy, galaxy formation, the history of science, etc. One problem, of course, is that no such person exists. The problem is amplified when we consider the expertise required to do justice to the subjects of the entire series: geology, biology, paleontology, ancient languages, archaeology, textual criticism, mideastern history, etc. Of course, I encourage the reader to read as many books as time permits in each of these fields. Skeptics, however, know that they must read a number of books in each field to get a rounded picture of it. "Authorities" in any one of these disciplines often contradict one another; and they find themselves competing with colleagues for attention and funding.

For this reason, unbelieving skeptics and Bible believers alike understand the need to hear more than one side of an argument. Everyone can identify with Proverbs 18:17: "The first to present his case seems right, till another comes forward and questions him."

And if specialists are often less than objective, their books (which use the language of the specialist) are often less than intelligible to the non-specialist. Some might go so far as to say they're boring.

This series gives me the opportunity to summarize the experts in plain English and to make an occasional outlandish observation of my own about the relevance of their findings to everyday life. Having brought all the relevant scholars to the table, I can then help the reader examine all sides of the issues. I claim to bring nothing to the table myself except healthy skepticism and a modicum of common sense.

For the sake of clarity and easy reference, I present the information under topical headings. Each chapter begins with a skeptic's questions and a brief summary of a Bible believer's response. I then give the evidence to support my position, along with (or followed by) an evaluation of alternative views. I try to help the reader explore the teachings of those in academic circles who have criticized the Bible, to get behind the broad statements and find out what kind of evidence these experts actually have to support their beliefs.

True scholars—those who devote themselves to astronomy, archaeology, textual criticism, etc., as a career—are supposed to use the inductive methods of science, meaning that they must begin their work simply by observing facts. If they start with foregone conclusions and then look for data to support them, their data will rightfully be suspect to the skeptic. Those of us who wish to make a defense of our faith (apologetics simply means "defense") are then permitted to point to the data of unbiased scientists to help us make our case. In the next four volumes, I try to distinguish between conclusions that are based on observations from science and those that are based on presuppositions of certain scientists (for example, I point out where both "creation scientists" and "macroevolutionists" are clearly biased).

In the spirit of skepticism, I try to restrict the premises of my arguments to the least we can reasonably know. Even while arguing from the least that can be said, it turns out that there's quite a bit that even a skeptic can put together, as shown by the conclusions in each chapter. I cite authorities—especially the better-known, unbelieving ones—to show that even those who are most opposed to a biblical viewpoint make amazing statements in support of it, particularly when they simply comment on the physical evidence. In the endnotes for each chapter, I usually support potentially controversial statements, not with quotes from Bible believers, but from unbelievers—sometimes even hostile witnesses to the facts.

Eta Carinae Nebula, NGC 3372, also called the "Keyhole Nebula," in the constellation of Carina. This bright, gaseous nebula shines upon us with light that has been traveling for 9,000 years. But nebulae like this one are only a tiny fraction of the distance from us as the nearest galaxies. —*National Optical Astronomy Observatories*

CHAPTER 1

A Word to the
Unbelieving Skeptic

*"But sir," Gideon replied, "if the LORD is with us, why
has all this happened to us? Where are all His wonders
that our fathers told us about when they said, 'Did not
the LORD bring us up out of Egypt?'"*
—*Judges 6:13*

The Healthy Skeptic

Some of the world's greatest people of faith started out as the world's
greatest skeptics. Note the leery questions and doubts of Abraham (Genesis
15:8, 17:17), Moses (Exodus 3:18, 4:1-13), Gideon (Judges 6:13-17, 36-
40), David (Psalm 22:1), Jeremiah (Jeremiah 12:1), all of Jesus' disciples—
most memorably, "doubting" Thomas (John 20:24-28)—and Paul (Acts
9:1-5), to name just a portion from a much longer list.

Notice how many of them asked something like, "How can I know
that this unbelievable thing you're telling me is true?" Abraham, the great
exemplar of faith, wanted *evidence* to assure him that the land would really
be given to him, as God had promised. Moses wanted sound reasons before
he could believe God's statement that the elders of Israel would even listen
to him. Gideon doubted that God could do the sort of wonders he had
heard about Him doing in the old days. All of Jesus' disciples initially
doubted that Jesus would actually rise from the dead as he had said he
would do. Most memorably, "doubting" Thomas said he wouldn't believe
until he had seen the risen Jesus with his own eyes, put his finger into the
nail holes and his hand into Jesus' side. And perhaps the most extreme case
is the apostle Paul, who once doubted the word of Jesus so strongly that he
devoted himself to persecuting all believers.

Had it been up to many of today's preachers, this kind of questioning
of God's word would have been dealt with by a swift lightning bolt.
Instead, our patient God's response was to provide skeptical inquirers with
the rational evidence they needed. The Lord gave Abraham the surest kind
of guarantee anyone could offer in his day—a covenant of blood. God gave
Moses a logical argument, several supernatural signs, and physical help from
his brother Aaron in order to assure him that God could indeed use him

despite his poor public speaking ability. God granted Gideon the unusual signs he asked for when he "put out the fleece."

Jesus gave Thomas the evidence he needed to believe, offering to let this doubter closely inspect him, to put his finger where the nails had been and his hand into the spear wound in his side. Jesus gave Paul proof of His divine power on the Damascus road and called him to start spreading the good news he had been quelling.

The Bible presents us with a God who is too big to be shaken by a little skepticism. Could it be that exercising skepticism is a normal, healthy part of using our God-given powers of reason? Could it be that the strongest faith is that which is put to the greatest tests? After such tests we may be in the best position to fully obey the greatest commandment—to love the Lord our God, not only with all our hearts and souls, but with all our *minds.*

All this is to say that, though many Christians have reserved the term "skeptic" for the most black-hearted among the heathen, skepticism can actually be a safer base to begin one's spiritual journey than an emotional experience or a desperate leap of blind faith. Sooner or later the questions we avoid will return, probably during the storms of life. The honest and sensible approach is to meet them head on, to ready ourselves with raincoats rather than to run around trying to dodge the raindrops.

The word "skeptic" is derived from a Greek word meaning "to examine." There is no consensus among philosophers about how to define skepticism today, but if we define a skeptic as one who personally examines every belief rather than leaving the thinking to others, we have a healthy skepticism that accords well with both science and the Bible. Such skepticism opens, rather than closes, the mind to the good news of Christ because getting to the truth becomes more important than proving one's side. Jesus said, "Everyone on the side of truth listens to me" (John 18:37). He said, "If anyone chooses to do God's will, he will find out whether my teaching comes from God or whether I speak on my own" (John 7:17).

In Acts 17:11, Luke commends the Bereans because they "examined the Scriptures every day to see if what Paul said was true." They exercised healthy skepticism, and the result was that many believed (17:12).

Keep Your Skepticism

After a person puts his faith in Christ, the Bible continues to encourage him to evaluate everything he is taught. The believer is counseled to test or examine all things (1 Thessalonians 5:21, 2 Corinthians 13:5, Lamentations 3:40). He is warned against allowing himself to be taken in by deceitful teachings (Ephesians 4:14), by human traditions (Colossians

2:8, Isaiah 29:13), or by his own emotions (Proverbs 28:26).*

But a healthy skeptic in today's information age doesn't limit himself to questioning his Bible teachers—he evaluates *all* the so-called "experts." He questions the words of politicians, media people, and even the self-assured scientists, taking nothing they say for granted. Their academic degrees can occasionally put them at a disadvantage when compared to a skeptic with common sense and a need for evidence. All too often an "expert" has devoted his life and intelligence to a particular school of thought, followed it without question, enjoyed mutual support from a closed group of colleagues—and never realized that some of his theory's grand and intricate constructs have been founded upon presuppositions that have little to do with truth.

An important creed of science is summed up by the old Latin maxim, *Nullius in verba:* "Don't take anyone's word for it." The history of science shows us that the greatest discoveries were made by people who questioned so-called "facts" that all the world thought were settled issues. The healthy skepticism taught in the Bible partly explains why so many of the greatest discoverers in science have been strongly committed Bible believers (see Bonus Section #1 for a listing of fifty of them and their achievements).

One might argue that if we become too rational, we leave no room for faith. However, I am not suggesting that we build our faith on rationalism, but merely that we build our lives on a faith that is rational. The evidences that point to the Bible's reliability and authority are first accepted by reason. Once a person is convinced that faith in Jesus Christ is unlike any other faith, that the Bible is grounded in history and is unique in its reliability, that person is faced with a choice to personally accept or reject its message. The choice to accept is called faith, but there is nothing irrational about it.

---

* The Bible tells its readers to look out for those who presume to speak in the Lord's name things He has not commanded them to say (Deuteronomy 18:20). Jeremiah speaks of those who prophesy "delusions of their own minds" (Jeremiah 14:14), Ezekiel warns of those who prophesy out of their own imagination" (Ezekiel 13:2), and Jesus told us to "Watch out for false prophets. They come to you in sheep's clothing" (Matthew 7:15). Paul said they "masquerade as servants of righteousness" (2 Corinthians 11:15), and John warns us, "Do not believe every spirit, but test the spirits to see whether they are from God, because many false prophets have gone out into the world" (1 John 4:1). Peter said that just as there were false prophets among the Old Testament people, so "there will be false teachers among you . . . . In their greed these teachers will exploit you with stories they have made up" (2 Peter 2:1,3). People who read their Bibles have no excuse if they're taken in by those cult leaders and healer-showmen who have never relieved folks of anything but their money.

Shocking Summary Statements
and
Stimulating Conversation Starters

◀ Some of the world's greatest people of faith started out as the
world's greatest skeptics, including Abraham, Moses, Jesus'
disciples, and Paul.

◀ The heroes of the Bible asked tough questions; God patiently
responded by providing them with the rational evidence they
needed.

◀ Skepticism can be a safer base to begin one's spiritual journey
than an emotional experience or a desperate leap of blind
faith.

◀ You aren't a real skeptic if you don't examine the evidence for
yourself.

Targeting Extraterrestrials . . . and Skeptics. The SETI (Search for Extraterrestrial Intelligence) Institute booked Australia's Parkes radio telescope, the largest in the southern hemisphere, for its targeted search for six months in 1995. NASA spent 60 million dollars on SETI. Now that it is privately funded, SETI's Project Phoenix plans to expand its tuning capabilities and target more carefully. What does this have to do with the next chapter? If scientists can go to such great lengths to target and tune into extraterrestrials, whose existence is questionable, perhaps the Christian witness could put as much effort into tuning into the skeptic next door, whose existence is certain. —*Courtesy of J. Masterson, CSIRO Division of Radiophysics*

CHAPTER 2

A Word to the Believing Witness

This is what is written: The Christ will suffer and rise from the dead on the third day, and repentance and forgiveness of sins will be preached in his name to all nations, beginning at Jerusalem. You are witnesses of these things.
(Luke 24:46-48) —Jesus

Tuning in to Skeptics

Seeing the title for this series, some Christians might at first react, "*Wonders* don't witness—*people* do." Facts from science and history seem too dry and impersonal to bring satisfaction to a thirsting soul. But the facts themselves do witness—if a skeptic is enough of a truth-seeker to dig into the evidence for himself. In actual practice, however, most people don't diligently seek the truth, and the Christian witness is necessary to draw a skeptic's attention to facts about which he has asked questions but never taken the time to search for answers.

The gospel must be *fact* if it is to be worth anything as a belief or a benefit. What did Jesus mean when He said, "You will be my witnesses" (Acts 1:8)? What are Christians to be witnesses *of*? Of the ineffable feelings of peace and contentment we experience when we know Christ? Any cult member might claim as much for his life-changing experiences. The disciples were witnesses of the sacrificial death and resurrection of Jesus Christ, without which the Bible says our faith is futile. Unless the gospel has changed, we must point to these same facts. And we must do so in the way that Jesus did, by first getting on the wavelength of the person we intend to reach.

Jesus consistently filled his teachings with common things easily understood by his hearers, beginning with their earthly concerns, not His heavenly ones. To farmers and fishermen He spoke of seeds and fishing nets. To the Samaritan woman at the well, He spoke of water. And to the skeptical Sadducees, who did not believe in the resurrection, He gave a powerful argument (about the living God of the still-living patriarchs) to show that their belief was inconsistent with the God of the Bible, Whom they claimed to worship. In each case, Jesus was ready to address the specific concerns of His hearers.

When Paul preached to the people of Athens, he started by quoting some of their own poets. He referred to the statements of respected writers in Greek culture in order to show a logical contradiction between these commonsense truths and their metaphysical beliefs. First he quoted Epimenides: "For in him we live and move and have our being," and then Cleanthes in his *Hymn to Zeus:* "We are his offspring" (Acts 17:28). "Since we are God's offspring," Paul argued, "we should not think that the divine being is like gold or silver or stone—an image made by man's design and skill" (v. 29). Before launching into his argument, once again, the evangelist was prepared to start talking about the views of his listeners.

One of the chief needs for the witnessing Christian today is to be prepared to reach skeptics. As this millennium draws to a close, American Christians are increasingly encountering people who lack even the most basic beliefs about the God of the Bible. Pastors, Bible study leaders, missionaries and concerned friends are often hard pressed to know how to begin to present the gospel to them.

George Barna's studies have shown that the percentage of Americans who describe themselves as skeptics has swelled in recent years. Skepticism has become a major identifying characteristic of "baby busters" (those now in their teens and twenties).[1] People from this age group characteristically question their parents' faith—and they question the *facts* behind that faith.[2]

An earlier generation actually led the trend away from belief in the Bible. In 1963, the Princeton Religion Research Center reported that 65 percent of Americans thought that the Bible is the actual Word of God to be interpreted literally. By 1994, this figure had dropped to 12 percent.[3]

George Barna's reports use a slightly different gauge to measure the changing beliefs. He uses the criteria of the National Association of Evangelicals to find that adults holding evangelical beliefs in 1994 comprised just 7 percent of the American population. This figure has dropped from 9 percent in 1993 and 12 percent in 1992.[4]

Many have used Gallup's surveys to try to show that belief in God and involvement in religion are as high as ever. After all, well over 90 percent of Americans believe in God and claim some sort of religious affiliation.[5] Recently, however, Billy Graham made a critical distinction: "While many Americans believe there is a God, most have not accepted true Christianity or Judaism or Islam. They believe the Bible, but they don't read it or they don't obey it."[6] We should remember that Paul started his sermon to the pagans of Athens with the words, "I see that in every way you are very religious" (Acts 17:22).

Nine out of ten Americans continue to identify with some religious group of faith. More than eight out of ten claim to be Christians.[7] But these figures say more about where Americans have been than about where they are. We have not considered the whole truth until we reckon with the fact that only 40 percent say they attended a church or synagogue in the past week, and some skeptical researchers suggest that about half of those who claim attendance are lying.[8] A greater number attend Easter services, but Gallup shows that many of the attendees do not know what Easter celebrates.

Surveys demonstrate that one in four adults cannot say what Easter's significance is; more than a third of younger Americans (age 18–29) do not know the correct answer.[9] And though 91 percent of the baby buster group say they believe in God, 12 percent say that God is the full realization of human potential, 8 percent say that God is a state of higher consciousness, and 4 percent said that every human being is his own god.[10] Only 25 percent identified themselves as born-again Christians.[11]

Obviously, true Christianity is more than church affiliation or general belief in a deity, and those who wish to share the good news of Jesus Christ should be aware that many who claim to belong to some religious group may have needs as great as the "non-religious." The fact that so many people claim religious affiliation makes the evangelist's job tougher, not easier. And the fact that most Americans have had some small exposure to the Scriptures without being grounded in them means that they may have just enough knowledge to recognize the apparent contradictions between science and faith. Their own limited faith may be shaken by the gentlest breeze; those who are fully exposed to the winds of naturalistic scientism are easily blown away. Many hold onto a wisp of faith only by concluding that truth is relative: the scientists have one truth, and the Christians have another truth that works for them.

What's the Most Difficult Thing About Witnessing?

For most of us, the most intimidating thing about sharing our faith is getting started. Unbelievers drift in and out of our lives and afterward we think, If only I'd said something while I had the chance. But all too often, the right chance never comes along.

Where do we start with someone who shows little spiritual interest or who shows skepticism toward our message? To follow Jesus' example, we can start with common beliefs and interests. If Jesus often started talking about seeds, harvests, grapevines, etc., in order to reach an agrarian society, then today, to reach people in an age when science has revolutionized their lives, we can use findings from science as an ice-breaker.

Many of this century's discoveries in science raise questions that science can't answer. In fact, many findings can't be explained apart from the God of the Bible. When you want to share the good news in a non-threatening way, a good starting place is to talk about some of these interesting discoveries that believers and unbelievers can agree upon. These findings can serve as a springboard to lead naturally to biblical answers and to the gospel of Jesus Christ.

Carl: I went door-to-door evangelizing once with my assistant pastor. It really got my heart pumping. That was about thirty years ago.

Fred: Did you see any results?

Carl: No, but there's a whole row of houses where the people have their blood on their own heads.

Fred: What was your opening line?

Carl: Hi, I'm Carl Klinkhammer. Do you know where you'd go if you were to drop dead right now?

Fred: That's pretty slick. The last time you tried this was thirty years ago?

Carl: Well, now I believe in friendship evangelism.

Fred: So how do you get into the subject of the gospel now?

Carl: Well, I haven't actually gotten too far into it with anyone yet. I'm mainly just concentrating on being a nice guy and getting along with them first.

Fred: For thirty years?

Carl: Well . . . come to think of it, most of my friends are Christians. I really don't have any opportunities to get to know non-Christians.

Fred: Really? Do you live in a neighborhood where everyone's a believer?

Carl: Well, it's hard to talk to *them*, except about recycling our garbage and the weather. One time I *almost* had an opportunity. My wife had invited a couple of neighborhood ladies to a woman's Christmas dinner at our church, and when they came back to our house, one of them wanted to know about what our church believed and so on.

Fred: That sounds like a great opportunity.

Carl: Yeah, well I think the devil was really attacking. My wife had laryngitis and I couldn't hang around right then because of my hair.

Fred: Because of your hair?

Carl: I'd just come in from shoveling the snow off the driveway. I'd worked up a good sweat, and so when I took off my ski hat, I looked in the mirror and parts of my hair were smooshed down and parts were standing straight up, you know? I looked like I was half-way through Hair Club for Men surgery.

Fred: So you couldn't talk to them because you were having a bad hair day.

Carl: My hair was a bad testimony. I probably would have done more harm than good, looking like that. By the time I got it straightened out, they'd all gone home. (Sigh) All right, maybe I blew that one. But the real problem is, opportunities like that hardly ever come up.

Fred: Maybe this book will give you a few ideas for ways to *create* opportunities.

Carl: You want me to be brutally honest?

Fred: Well, honest, anyway.

Carl: The truth is, I get panicky just thinking about trying to share my faith with unbelievers. I mean, here I am, with their eternal destiny in my hands, and I know I'm going to come off sounding like a fanatic—and that's not going to help them. I mean, I know the Holy Spirit does the convincing, but I also know that, like you said, we're living in a skeptical age. A lot of these people don't even believe in a personal God. I know I'll sound like a real jerk to them when I start talking about sin and salvation.

Fred: You don't have to start right out talking about sin and salvation. In fact, I'd advise against it, especially if the person is still hung up about whether there's a personal God.

Carl: But the *goal* is to get the person to pray the sinner's prayer, right?

Fred: Not necessarily the first time you talk to him. If you try to skip right to the sinner's prayer before the person has even recognized that a personal God exists, you may have overlooked a significant step for that person, to say the least. Hebrews 11:6 says that anyone who approaches God "must believe that He exists and that He rewards those who earnestly seek him." This is why the subject of this book is relevant when you're trying to figure out how to start sharing your faith with people who are skeptics.

Carl: So what's the perfect conversation starter?

Fred: Well, here's *one* way to start. Ask a non-threatening question from science. Tell your friend you've been reading a book about the ori-

gin of the universe and ask what he thinks about the big bang theory or the fine-tuning of the universe. Or pick any of the other
seven wonders of this series and give him a chance to talk about
what he thinks about it. Most people like to give their opinions,
whether or not they know anything about science. But it gets them
thinking about ultimate causes and it gives you the right to respond
with your thoughts.

Carl: And *then* you sock them with the four spiritual laws.

Fred: If you're talking to a skeptic, he may not be ready for that yet.
Don't get me wrong. If the person's ready, by all means, you want
to be ready to share the good news about what Jesus came to do.

Carl: My Bible's been marked to take someone down the Romans Road
for 30 years.

Fred: The goal is to talk to the person often enough so that he's ready
for that. It might help to picture a person's spiritual journey as if it
were the face of a clock. Think of twelve o'clock as the time when
they hear God's call and they're ready to recognize their sin problem and the Bible's solution. If you run into someone who's right
there, say at eleven o'clock, then by all means, help him follow
through. But if someone's at one o'clock, you should be prepared
to help take him around the clock, hour by hour.

Carl: So for the guy who's obviously at one o'clock, you're saying I can
start asking what he thinks about the origin of the universe or
about its fine-tuning.

Fred: Yes, then you can talk about the findings of science on the subject,
in an unbiased way. You know, here's what Einstein said, here's
what Hubble said. Talk about the questions science raises that it
can't answer. And then you can say how you find it interesting that
the scientific findings fit what the Bible has said all along; you can
say that it looks to you like the God of the Bible provides the super-
intelligent design and the care that science points to.

Carl: And *then* you can start in about sin and salvation.

Fred: If you sense the person has come to agree that the Bible is credible, then you might be ready to talk about the most important
thing the Bible has to say: the good news of Jesus Christ. But this
may not happen in one discussion. The best situation is one
where you can get the person to say he wants to hear more and
to commit to having regular discussions with you about it. For
people who are starting with very little Bible knowledge, you can
ask them if they'd like to go through a survey of the Bible with

you.* You might be surprised to find how many people are open to learning more about what the Bible has to say, simply as an educational benefit. If you're willing to prepare well and take one person through a series of weekly studies on the Bible, you might be surprised to find how confident you'll become and how easy it is to share God's Word with more people after that. It can be habit-forming.

Why Are People More Skeptical About the Gospel Than Ever Before?

Young people—both believers and unbelievers—take science courses and are given the impression that evolution and other theories of science can explain the natural world, making the Bible appear unnecessary or even contrary to reason. Many lose interest in our faith when their questions go unanswered. Others keep the faith the best they can, but become adults who harbor fears that their faith is groundless; and so they never give the Lord their all.

Parents, youth workers, Bible study leaders, and pastors need to get ready to tackle the scientific problems raised by the schools and by the media. Gaining an understanding of the bare evidence in science actually gives Christians the opportunity to turn the situation completely around: the Christian can raise far greater problems from science for unbelievers than unbelievers can raise for Christians. At least the Bible offers logical explanations. Science has *no* explanation for the ultimate origin of the universe or for its incredibly precise fine-tuning for our benefit. But we can't expect people to take our word for these evidences from science; we need to use the words of the unbelieving scientists themselves. One purpose of these books is to equip the concerned reader to do just that.

The Gospel for Grownups

In order to reach adults today, many of us are continuing to use evangelistic strategies that were developed to reach teenagers decades ago. Is it any wonder that post-adolescents do not become Christians as frequently as adolescents do? At the beginning of the media's current interest in our

---

*Searchlight plans to print a simple survey of the Bible as a leader's guide to help Christians take their friends through the Bible in four or five sessions. In the meantime, here are some other books that might be used for this purpose: *What the Bible Teaches*, edited by William MacDonald (Dubuque, IA: Emmaus Bible College); *What the Bible Is All About*, by Henrietta Mears, and its accompanying *Group Study Guide*, edited by Wes Haystead (Ventura, CA: Gospel Light); *Unger's Survey of the Bible*, by Merrill F. Unger (Eugene, OR: Harvest Bible House); *All About the Bible*, by Sidney Collett (Uhrichsville, OH: Barbour); *Books and Parchments*, by F.F. Bruce (Grand Rapids: Fleming H. Revell Co. Div. of Baker Book House); *The Bible from Scratch*, by Simon Jenkins (Batavia, IL: Lion); *Message of the Bible*, edited by George Carey (Batavia, IL: Lion); and *Compact Survey of the Bible*, edited by John Balchin (Minneapolis: Bethany).

age of skepticism, *Christianity Today* carried Larry Poston's insightful article, "The Adult Gospel," which pointed out that the average age at which
people in the West convert to Christianity is 16, while the average age of a
Westerner's conversion to Islam is 31.[12] Poston concludes that "Muslims
are apparently reaching an age group in Western societies that Christians are
not."[13] Christian faith is often perceived as requiring an irrational leap, while
Islam is presented in a way that appears simpler and more reasonable.

Getting on the same wavelength of many people today means that we
recognize that they are skeptical about the Bible's claims. Because argument
by an emotional or prejudicial appeal (ad hominem) is apparently losing its
effectiveness in this age, the arguments in this series are made "*ad scep-
ticum*"—they are meant to appeal to the skeptical nature, which is among
the highest and most useful qualities of the human intellect. Putting ourselves into the other person's shoes, we can understand why most would
rather be appealed to on the basis of their higher rather than their lower
qualities.

Getting on the same wavelength of many people today means recognizing that they may have a higher regard for science than for preaching.
And they're more interested in self-improvement and in *expanding* their
perspective of spirituality than in narrowing their views into denomination-
al categories.

Independent-minded adults are not motivated to make life-changing,
spiritual decisions by an opportunity to join another group—at least not to
the same degree that teenagers are. The idea that our adult acquaintances
will become Christians when we invite them to church, that they just need
to be exposed to the gospel by hearing our pastor, is an idea that is questionable for at least two reasons. First, it's not as likely that we'll get an adult
to give up his Sunday morning to come with us. And second, even if he or
she comes, an adult's attendance at a church service does not have the same
impact as a teenager's visit to a youth group. For an adult, a one-on-one or
small group discussion—between friends—can have far greater impact, if it
is done in a way that deals with the real issues of his concern. These books
are intended to be a tool to help address some of those concerns.

Society has moved on while many Christians have remained stuck in
the rut of their own little cultural enclaves, some using overly-emotional
preaching styles and irrational (and often unscriptural) practices that ensure
a narrow hearing. Worse, many are even broadcasting their denomination-
al excesses, somehow finding the greatest charlatans in all Christendom to
be their spokesmen. Money spent on this "evangelistic" outreach, of
course, goes to reaching an almost exclusively Christian audience. Christian
TV, originally founded for the purpose of spreading the gospel, now has as
its purpose the hawking of "Jesus junk." As one president of a marketing

company told me, "Christian TV is now direct response marketing in its purest form." Those unbelievers who do catch a glimpse of "Christianity" while switching stations are thus more firmly convinced that the Christian faith is indeed irrational. It's for unthinking herd-followers.

Perhaps the best thing a Christian witness can do first is to demonstrate that we—and what we have to say—are *not* irrational. We have not turned our backs on science; rather, Christians have led the way (again, see Bonus Section #1). We are not unthinking "flakes" who mindlessly follow the traditions of our particular church. We are conscientious people who, having thought through the issues, have decided to commit our lives to the one true God, the Creator of the universe, as revealed in the one book that gives evidence of His inspiration.

Once we have convinced someone of this, once he no longer feels threatened by our flakiness, half the battle is won, because a reasonable person is usually open to the truth. Even if a person doesn't particularly *feel* that something is missing from his life, he is usually interested in increasing his knowledge of the Bible as one of the world's great books.

At no time in modern history has the average Westerner been so lacking in the foundational knowledge of the God of the Bible, of the events the Bible describes, and of Jesus' place in history.* Getting on the same wavelength of people today often means that we recognize that they lack the foundation for belief that past generations had. To us, laying that foundation is the joy of witnessing. From our *friend's* perspective, we are meeting an educational need—whether through informal discussions or an organized, weekly neighborhood Bible study. This type of evangelism may be a more realistic means of leading people to Christ today than the type of evangelism that tries to do it all in one sudden blast from our gospel gun.

There will be some cases in which the other person's wavelength is irrational and not a good starting place for rational discussion. Many people today have been influenced by Eastern religions, New Age thinking, or just plain subjectivism: the idea that your truth is fine for you and my truth is fine for me. They may seem impenetrable. There is nothing reasonable about believing that historical facts may vary from person to person.

But even in these cases we should show the courtesy of hearing out the other person's views. The concept of truth's relativity sounds broadminded only until they face the logical contradictions that must result. While many find it pleasant to believe that there are many roads to heaven,

---

*Agreeing with George Gallup, Jeffrey Shelter (Religion editor of *U.S. News & World Report*) says that America has become "a nation of biblical illiterates."[14] Charles Colson claims that we are living in a post-Christian culture; he writes that "we are in much the same situation as the first-century church, needing to educate in order to witness."[15]

most will recognize the fact that a person cannot go to heaven and come back reincarnated as a cow at the same time. Encouraging them to clarify their beliefs with concrete examples (of how truth can be relative) can sometimes be an argument in itself. Having then earned equal time, we can still draw the person's attention to the facts presented in this series from science and history, hoping that in their inmost mind, they will recognize the voice of reason.

The Most Convincing Arguments

Of course, the most convincing arguments may be meaningless to a skeptic if he is really looking to see what difference Christ has made in our lives—and he sees nothing. By the same token, a changed life may also have no impact if we're not ready to talk about the explanation for it.

When we speak of presenting an "argument," of course, we don't mean we engage in an emotional quarrel; we mean we present a conclusion properly related to its supporting evidence. Until we present the evidence that leads to our conclusion, we have no argument—we have an unsupported opinion, which may only succeed in giving others supporting evidence for the conclusion that we're pompous blowhards. Proverbs 18:2 says, "A fool finds no pleasure in understanding but delights in airing his own opinions."

The evidence we present to support our conclusions is most effective if we begin with premises our skeptical friends already hold to be true. If we're talking about cosmology, the authorities we cite most often should be respected for their renown in science—not theology. As you read through this book, I suggest that you keep a writing pad handy to make an occasional note. Look for creative ideas of your own about ways to start a profitable conversation with a skeptic.

1 Peter 3:15 tells us, "Always be prepared to give an answer to everyone who asks you to give the reason for the hope that you have. But do this with gentleness and respect . . . ." The fact that most contestants in the debates of our audiotape set are violently catapulted from the Grand Skeptic's "Refuge of Reason" is not meant to encourage us to thoughtlessly cast out other people and their ideas when they don't agree with us. It is meant rather to help us picture what others may do with *our* views when they ask us for *reasons* for our hope and we give them only opinions. How long we remain in the realm of their serious consideration depends on whether we're ready to give reasons, whether those reasons are reasonable to *them*, and whether we present the reasons "with gentleness and respect."

Shocking Summary Statements
and
Stimulating Conversation Starters

Today, most Westerners would claim to believe in God and be affiliated with a church—yet most seriously question the basis for their parents' faith. To get on their wavelength, the Christian witness must understand their skepticism.

Waiting for the right chance to share your faith is a tried-and-true way to avoid having to share your faith.

A good way to begin to present the gospel to a skeptic is to talk about scientific discoveries that both believers and unbelievers can agree upon.

Those who control the media (including Christians who control the Christian media) often present our faith in a way that makes it appear irrational. The first responsibility of the Christian witness is to counteract this notion by showing that he and what he believes are *not* irrational.

A true skeptic will appreciate the opportunity to learn more about the Bible as an educational benefit. Taking a friend through a five-session survey of the Bible is a non-threatening way to lay a solid foundation for the gospel.

Notes for Chapter Two

1. George Barna, *The Invisible Generation: Baby Busters* (Glendale, CA: Barna Research Group, 1992), pp. 31, 44.

2. Ibid. *Time's* July 16, 1990 cover story asks the question, "Why are today's young adults so skeptical?" The article typifies the media's general recognition, beginning especially with the 1990s, that we have entered a new age of skepticism.

3. "20 Questions About Religion," *PRRC Emerging Trends* (Journal of the Princeton Religion Research Center), (June 1994), p. 4.

4. "Barna Report Indicates Decline in Evangelical Base," *Bookstore Journal* (November 1994), p. 22.

5. "Religion Index Hits All-Time Low Mark," *PRRC Emerging Trends* (March 1994), p. 2.

6. Billy Graham, quoted by John McLaughlin on John McLaughlin's *One on One*, taped April 1, 1994, for PBS television.

7. *Gallup Poll Monthly*, (September 1993), p. 27.

8. "Do That Many People Really Attend Worship Services?," *PRRC Emerging Trends* (May 1994), p. 1.

9. *Gallup Poll Monthly* (March 1991), p. 57.

10. Barna, p. 157.

11. Ibid., p. 31.

12. Larry Poston, "The Adult Gospel," *Christianity Today* (August 20, 1990), p. 24. Larry Poston serves as chairman of Nyack College's Department of Missiology in Nyack, New York.

13. Ibid.

14. Jeffrey Shelter, in interview for John McLaughlin's *One on One*, taped April 1, 1994, for PBS television.

15. Charles Colson with Ellen Santilli Vaughn, *The Body—Being Light in Darkness* (Dallas: Word Publishing, 1992), p. 332.

PART II

A Little Science Fiction

(Before the Science Facts)

Carl: I like the idea of your putting in some fiction here. Religious fiction is very hot these days. But *science* fiction will cut out over half your audience. Science fiction only appeals to men.

Fred: That's gender stereotyping.

Carl: That's marketing truth.

Fred: What if there's a love story in it?

Carl: *That's* gender stereotyping. But a little romance will definitely help with the women.

Fred: Before we get real deep into the science, I just want people to see how these issues apply to our own, personal lives.

Carl: You mean like, everybody wants to be more attractive to the opposite sex, and greater knowledge of theoretical physics will make you more mysterious?

Fred: Uh, no. I mean like, everybody wishes he could actually see God or hear directly from Him, to be sure that He's really there, to know whether He's involved with us personally. And science gives us one way to see God, or at least to learn a lot about him. But just *living* gives us another way.

Carl: Still, I wish we could go back to the old days, when God would actually show up and talk to those Old Testament people.

Fred: Actually, the "hiddenness" of God is a major theme of many of the Old Testament prophets. God reveals Himself through nature, through our consciences, and finally through His Word, but I don't believe that God makes it a regular practice to reveal Himself *directly* to most mortals—ever since sin entered the world. Sin may be the main reason. In this short story, I give another reason.

Carl: Because many of us would have heart attacks?

Fred: No, it has to do with having the opportunity to trust God without being forced to do so. But even if God is hidden, in one sense, in another sense He's very hard to avoid.

Carl: You mean for scientists?

Fred: I mean for everyone. There was a time in my own life when I seriously questioned God's existence. After I'd had what I considered to be a particularly lousy ten-year stretch, I went to see a pastor friend of mine. To sum up my side of the conversation, I basically told him, "I no longer believe in God. Do you think that's why He's so upset with me?"

Carl: We need a snare and drumbeat there.

Fred: God is hard to avoid.

Carl: I can see how God is hard to avoid, on an emotional level, especially for someone who was brought up as a Christian. But what about for non-Christians?

Fred: Part of the point of this story is that God is hard to avoid, emotionally or intellectually, for *any* thinking person.

Carl: Especially if they're getting a message from space, huh?

The Arecibo Observatory in Puerto Rico, the world's largest radio telescope. The SETI Institute's Project Phoenix uses this 1,000-foot colossus for its revived and targeted search for extraterrestrial intelligence. This much is not fiction. —*Courtesy of Arecibo Observatory*

CHAPTER 3

The Message from Space

or

Doing Research for Fun and Profit

Why spend money on what is not bread,
and your labor on what does not satisfy?
—*Isaiah 55:2a*

THE GREATEST DISCOVERY of all time was made by one who had neither earned nor expected it. After NASA had spent sixty million dollars and the SETI Institute folks had spent many thousands of man-hours in their search for extraterrestrial intelligence, a seven-year-old girl independently detected the first signal from an intelligent source outside our solar system. Some, already convinced that the cosmos was godless, believed that this provided additional evidence that there was no justice in the universe.

The first encoded message came in the form of a series of flashes from the vicinity of Epsilon Eridani, a sun-like star that had been observed as a good candidate for intelligent life, but which had long ago been dropped from consideration because of its radio silence. The radio astronomers of the SETI Institute re-examined it when the flashes of light were brought to their attention by a second grade teacher in Pittsburgh. Her class had been using California's Mount Wilson observatory by remote. The SETI astronomers recognized the flashes as a beacon to draw their attention to the more detailed microwave message emanating from that location. Thus the greatest discovery of all time became known as "Gracie's message," after the original signal's seven-year-old discoverer.

But our story concerns the man who ended up receiving the bulk of the credit for this greatest of all discoveries, the man who almost single-handedly found the funding to bring the search for extraterrestrial intelligence into the 21st century after the 20th century had about given up. A relative latecomer to the cause, Mark Rizley had no training in astronomy or physics, but his motivation to do something important with his life had begun ten years before the discovery, on a Monday morning.

*   *   *

Mark reached for the switch to turn off his music alarm and lay back to ask himself what he asked himself every morning: Now what do I have to do today?

Before he had quite decided whether his apprehensions about his morning classes were well grounded, Margaret interrupted him with a thought that seemed to come from a million light-years away. "I think I've figured it out," she announced. Mark turned to see her staring up at the ceiling as if she had been awake for a while. This was unusual. Usually he didn't see her awake until after lunch.

"What? You've figured what out?"

"What life is," said Margaret.

She probably used the long pause that followed for dramatic effect, but Mark used it to decide that he should skip one of his morning classes in order to better prepare for his afternoon test on buyouts and divestitures.

"Life is a big research project," said Margaret, still staring at the ceiling. "We've been put here to do research to find out why we've been put here."

She's been a little weird ever since her father died, thought Mark. Either that, or this is what comes of all those philosophical bull sessions she has with the eggheads in the astrophysics department.

"Do you know a better reason for our being here?"

Mark thought for a moment and finally grunted.

"Research," said Margaret. "That's our real job. Everything else we're doing is incidental."

"Mm. Well . . . I'm going to be late if I don't get going."

"I mean, are we living just so we can go to school, so we can work . . . so we can have food and shelter . . . so we can live . . . so we can work?"

Mark slowly nodded. He tried not to seem impatient. After all, he could never take Margaret for granted. This was the relationship he had waited for all his life.

"Or," she said, while sneaking her arm around his waist, "are we living just for love?"

"I think that's it. But it's 7:05."

Margaret became philosophical again: "Or, if you look ahead a little, is our whole purpose just to create as loving and as successful a family as possible?" Mark was thinking now. Margaret looked at the clock and said, "You'd better get going."

When Mark came out of the shower, Margaret asked him, "So what do you think?"

"About what?"

"About what we were just talking about?"

Mark's shower had been spent planning his day. He changed gears quickly and said, "I think you're onto something. I mean, you're thinking about what's important. Some people just go through school on auto-pilot. But we know *why* we're studying and working so hard. We have purpose behind it. We're—like you said—we're planning ahead for a family, a successful family, and that takes a successful career."

Margaret apparently had nothing to add. Mark put on his glasses to see what kind of expression she'd been wearing during his little speech. She was staring ahead like a zombie. Sometimes Mark wondered if the differences between them were just a male-female thing or if no one really lived behind those beautiful eyes. Then she looked back at Mark and he saw her soul return.

"You've completely missed what I'm talking about." There was fire in her eyes now too. "I'm not talking about making a living—I'm talking about *living*."

So what do I have to be to know about living, thought Mark, an astrophysicist? He decided he would only make things worse if he said anything more, but Margaret took his silence for ignorance.

"You don't have a clue about what I'm talking about, do you?"

Mark felt as if she had just stabbed him in the stomach. Maybe he just needed breakfast. But here he was, thinking she was a zombie, and all along she was thinking *he* was a zombie, thinking he lacked the soul that allowed her to ask her lofty questions. He wondered if he had just blown all his months of courting, all his efforts at making good impressions. As he walked to his first class, he vowed that someday he would make something of himself that would show her that he was not a man to be lightly esteemed. He would be a mover and a shaker in the world while all her physicist friends would be scrambling for obscure teaching and research jobs.

*   *   *

Five years later Mark woke up in the middle of the night and asked himself, Now what do I have to do? Margaret was gone and both his children were crying. He heard Margaret groaning in the bathroom. Uh-oh, thought Mark. She only does that when she's throwing up. He walked into the children's room with a flashlight and confirmed the worst. Both kids had thrown up too. This would mean he'd have to get out the bucket, the rags, the disinfectant; he'd have to scoop up the vomit, carry it down to the laundry room sink, change everyone's pajamas and sheets and scrub the carpet—and then there was likely to be a repeat performance. It was going to be a long night.

Two hours later, during the third round of clean-ups, Mark opened up the children's window and inhaled a breath of fresh air. Margaret held the baby in a rocking chair beside him and apparently both of them had finally fallen asleep, but his three-year-old son was wide awake. The toddler was trying to get his dad's attention to point out a spot on his bed that had been missed. "Why am I doing this?" Mark mumbled into the night.

"Research," said Margaret.

*   *   *

Three years later Mark looked out another window at three in the morning. He couldn't stop reviewing his current state of affairs. His business had failed. His trusted partner had cheated him. He was in serious debt. He would have liked to have asked Margaret if this was all part of life's research project, but Margaret and the children had left him a month before.

Over and over again, Mark presented his case to an imaginary judge, explaining all his actions, making sure that all his friends and family in his imaginary courtroom understood his side of things. Sometimes, if matters got too personal, he would bring his problems to an imaginary, very understanding psychiatrist instead; but neither the judge nor the counselor could help him feel quite fully absolved.

And so, out of desperation for sleep, he looked up to the stars and tried talking to God: God, do you see the situation here? Is this something you can do something about, or don't you get involved in individual cases? If you do, if you could just put things back together between Margaret and me, and if you could get me back on a good financial footing again, I'd . . . I don't know—what do you want out of me?

Mark waited a moment and then asked out loud: "Are you there, or am I talking to myself?" He actually hoped for some kind of answer—perhaps a sign among the stars, but the night was still. "Of course, if you *aren't* there, then my situation doesn't really matter to anyone but me . . . and *I* don't matter. Nothing matters. Oh God, I hope you're there. Could you show me maybe a little sign? Well, actually, I guess it would have to be a *big* one if I'm really going to know it's you, if it's not too much trouble." The stars were motionless. "All right then, how about a *little* sign?" For the next half hour, he searched the sky, but absolutely nothing happened.

Mark decided that, if there is a God, He is as cold as outer space, as silent as the moon.

*   *   *

Waking up a few years later in a luxurious suite in the largest palace in the world, Mark wondered if his past troubles had merely been a bad dream. Humbling himself before God had been part of the nightmare. Margaret opened a window so they could hear the muezzin's lofty call to prayer. The couple were guests of the Sultan of Brunei, the world's wealthiest man, and they were staying in one of the 1,788 rooms of his four hundred million-dollar home.

Mark's financial problems had been solved by his success as a grant proposal writer, finding funds for various scientific research projects. Margaret, who had returned to him after two months of separation, had a never-ending list of friends in universities around the country in need of program funding. His knowledge of finance, combined with her knowledge of science, had made the team a hit in the grant proposal writing business.

Because they based their fees on a percentage of grants received, and because they had been successful in obtaining a number of sizable grants, they were well on their way to achieving the American dream of financial independence. Not bad, considering that Margaret's contribution was only part-time while she worked full-time with the team of scientists at the SETI Institute near San Francisco. However, Mark's optimistic financial projections depended upon his continuing to find grant money at his present rate; there was always the chance, he feared, that he had just been lucky so far.

Lately, Mark's biggest scores had been for the SETI Institute itself. Margaret's favorite cause had been in desperate need of capital ever since Congress had terminated all funding of SETI in the mid-1990s, leaving the Institute to fend for itself. Politicians had become anxious to show that they were more concerned with deficit reduction than with finding evidence of extraterrestrial intelligence. Mark had brought in needed cash from private foundations just when enthusiasm from SETI's former supporters had about run out. The Sultan of Brunei warranted a special trip because the Sultan himself had initiated contact, expressing his interest in SETI's "Project Phoenix," the targeted search that had arisen from the ashes of NASA's old program.

Over a catered breakfast in their suite, Mark and Margaret discussed strategy for soliciting what they hoped would be their largest contribution yet. For Margaret, this would save SETI, and for Mark, it would bring them into a new tax bracket.

"Be sure to emphasize the time crunch," said Margaret. "In a decade, interference from telecommunications satellites will clog the microwave bands. Tell him it's now or never."

"I think you should do the talking whenever it gets into the physics."

"You're the expert on selling it, though."

"Well," said Mark with a mouth-full of prawn, "the main thing we've got to let him know is how big this is. I mean, here's a man who's given half a million to New York's elderly shut-ins and a million to UNICEF. If we want to score that big or bigger, we've got to convince him that this project is at least as big as those causes. Here's what I want to tell him," said Mark. "Your Excellency—"

"Is it 'Your Excellency' or 'Your Majesty'?"

"I think either one's okay. Now tell me if you think this is too much." Mark cleared his throat and put down his fork. "Your Excellency, you could have a leading role in what everyone agrees would be the greatest discovery of all time. That's why donors are offering million dollar grants. Intel's chairman Gordon Moore says, 'I can hardly think of a more profound question to ask than, "Are we alone in the universe?"'

"And yet," continued Mark, becoming still more intense, "there's a much better reason for pursuing this than just to satisfy our curiosity. Our civilization has only been around for a few thousand years out of the ten or twenty billion since the universe began; we're newcomers on the scene. Anyone else we find out there is bound to have been around a lot longer than we have. Contact with an advanced culture would be science's most profound contribution to our civilization. They might tell us how to solve the problem of disease, of crime, of war. A civilization that survives for millions of years would have worked out solutions to these problems long ago."

Mark drank in Margaret's look of admiration. Now, he thought, she can no longer say my work is incidental to the real purpose in life. What could be loftier than this?

When the telephone rang, Mark and Margaret stared at the wooden pony where the ringing came from, realizing for the first time that it was a telephone. Mark pressed a button, activating the speaker-phone. "Hello?" said Mark.

"Good morning, Mr. and Mrs. Rizley," said a voice in crisp, Oxford English. "I trust you found your accommodations adequate last night?"

"Yes, wonderful," said Mark.

"I've been reading over the literature you sent me. It's fascinating." Each time the voice spoke, the pony's mouth kept opening and closing.

Margaret started to laugh, but Mark shushed her. He realized that they were speaking with the Sultan himself.

"Now as I understand it," said the Sultan, "your operating budget requires three million just to keep the search going another year."

"That's right," said Mark, talking to the pony.

"And you're looking for another million dollars to double the seventy-four million channels you're now receiving."

"Yes."

"But I also understand that you've doubled the spectrometer once before, thinking that this would provide you all you need to find a signal—and still you've found nothing."

Mark and Margaret looked at each other. The Sultan knew more about Project Phoenix than their literature had told him. "In the long run," said Mark, "doubling the spectrometer saves money by halving the search time."

"And if you *don't* ever find signals from an intelligence," continued the Sultan, "have you seriously considered the possibility that all this money and research will be completely wasted?"

Margaret took up the challenge and talked to the pony: "Even if we find nothing, the technology that's developed will be applied to medical diagnostic imaging, materials testing—"

"Of course if you don't find any other civilizations out there," interrupted the Sultan, "*that* would be an answer too, wouldn't it?"

Mark and Margaret weren't sure how to respond.

"I'm looking forward to meeting you this morning," said the Sultan, "but I'm curious about one more thing now: How will you be sure when you've received a signal from an intelligent source? How do you tell the difference between a natural, pulsating signal, as from a pulsar, and a signal from an intelligence?"

"If you get a message in Morse code," explained Margaret, "you know there's intelligence behind it. Nature can't duplicate that. In the same way, if we get a signal containing encoded information, even if we can't break the code at first, we'll know it's coming from an intelligent source."

"You mean like the encoded information we find in DNA."

Margaret didn't hear this interruption and continued her speech: "Nature can't duplicate specified complexity. The chance of nature creating a pattern that has meaning is almost infinitely small."

"Like the specified complexity of hemoglobin."

"I beg your pardon?" said Margaret.

"You know," began the Sultan, "it occurs to me that perhaps—just perhaps—your investigations have already given you your answer, if you are prepared to accept it. What if this encoded information you are seeking—

this signal of extraterrestrial intelligence—has been coming through loud and clear all along, but you've been missing it, because you've been looking in the wrong place?"

"Well, that's why we're continually looking in new sections of the sky—"

"Have you looked into hemoglobin? This is very interesting. I was just reading about hemoglobin. Are you familiar with the calculations of Hoyle and Wickramasinghe? I have it right here."

After a pause, they heard the Sultan flipping pages. "Here," said the Sultan. "Did you know that the chance that amino acids would line up randomly to create the first hemoglobin protein is one in ten to the 850th power? Which is in the realm of infinity, considering the fact that there are only 10 to the 80th power atoms in the entire universe."

Mark and Margaret could only stare at each other. This man obviously had more than a casual interest in science.

The Sultan continued: "There's an even smaller chance that the DNA code could have randomly reached the required specificity: one chance in 10 to the 78,000th power even for the DNA of a simple microorganism. *This* must be a signal of intelligence, wouldn't you say?"

Mark began to feel like the human prey in the story, "The Most Dangerous Game." Perhaps the Sultan had invited them there just to play games with them.

The Sultan closed the conversation. "I'll have a driver pick you up at ten, and we can continue our discussion at the polo farm, all right?"

A servant guided them through the long, echoing halls. Mark could see how it would be easy to get lost in a palace that covered fifty acres. The couple joked with one another about talking with Mr. Ed and getting their information right from the horse's mouth, but neither felt as confident about the success of their mission to Brunei as they had an hour earlier.

After the driver had given them a tour of the grounds and the air-conditioned stables, a servant seated Mark and Margaret at a shaded table near one of the polo fields. The handsome Sultan of Brunei, the 29th of his line, rode up to them on a white Argentine pony and dismounted. He looked fiftyish but his face was well-seasoned, judicious-looking, as a ruler's face should look, Mark thought. The monarch took off his riding cap and exchanged greetings with the couple. After sitting down with them, he seemed content to study them until they were ready to start the conversation.

Mark hurried to think of a way to start with some small talk. "I don't think even *Solomon's* stables were air-conditioned, were they?" he said.

"The ponies are my passion—just as, I am sure, SETI is your passion." The Sultan searched their eyes. "It is a pity your government does not share your passion, hmm?" And then he turned to Mark. "Mr. Rizley, how passionate are you personally about Project Phoenix? What would you be willing to give up in order to solve the riddle of the universe?"

"Well, to be frank, Your Excellency, it's always seemed to me that the smartest plan is to find a way to do these things without giving up anything."

The Sultan did not miss the glare Margaret shot at Mark. Mark realized that his worldly-wise answer had not impressed either of them. As if to atone for himself, he launched into his how-to-solve-the-disease-crime-war-problem speech.

The monarch listened politely with his chin resting on his folded hands. After Mark was through, the Sultan gazed out over his polo fields and said: "Let's look at the possibility that you continue to find nothing. Would that prove that we are unique as creatures in the universe?"

"Well," began Margaret, "I suppose that would tend to show that life-forming processes might not be as inevitable as we think. But I believe that negative results are unlikely. We wouldn't be doing this if we didn't think our chances were excellent of finding something—in this decade."

"Because you believe life-forming processes are inevitable," said the Sultan.

Mark wondered, Is that the Sultan's angle? Is he interested in this only because he wants the results to be negative? Is he hoping to score points for special creation or Islam? Mark decided he'd better be careful not to say anything that might needlessly discourage him.

"You raise an interesting point," said Mark. "If the results are negative, creationists could probably make a lot out of that. You could interpret the evidence to say that we're unique."

"But the real point is," said Margaret, who had apparently missed the real point, "most scientists today believe that the universe must be humming with life—all over the universe. If we can double the channels and can work on this another five years, we're bound to find other civilizations that are continually sending out incidental microwave radiation, like we do."

"But you're right," added Mark to the Sultan, "if the results are negative, *that* would further our knowledge too."

"It would make us . . . hard to explain, wouldn't it?" This time he searched Margaret's face, but she met his eyes coldly. "Now," continued the Sultan, "let's look at the other side. If you *do* actually find a signal from an intelligent source, what do you think will be the impact upon the world?

What will be the impact upon religion, for example?"

"Well," said Mark, "it would just show that Allah created more intelligent beings than the ones on this planet."

"I think Mrs. Rizley knows better. You are expecting to find life because you think we are the product of nothing special. Just biological inevitability. The playing out of nonsense laws that created sense, as some of us have in small quantities. If we are here, you reason, it is only because it was biologically inevitable. And if the right conditions exist elsewhere, it is inevitable that life will emerge and evolve there also.

"The right conditions do exist elsewhere," insisted Margaret. "There are a half billion sun-like stars in this galaxy alone and we now have evidence that many of them have planetary systems—"

"You may be right—about the conditions. But that is not the issue. Is it inevitable that life will arise whenever the 'right conditions' exist?"

"That's what we want to find out," said Margaret.

"You mean that's what you want to prove. I'm the one who wants to find out." The Sultan scooted his chair back from the table and squinted up to admire a mountainous cloudscape. "You have both bargained very well today. I will give you one million U.S. dollars toward operating expenses— and another million to double the spectrometer." Then he looked back at them. "Because I want to know . . . I really want to know." He turned to the driver standing by a nearby gate and said something to him in Malay. The man opened the gate for them and Mark realized the interview was over.

"Thank you, Your Excellency," said Mark. "Thank you very much."

As the Sultan mounted his pony, his eyes caught Mark's one last time. "If you were me, Mr. Rizley, I don't think you would give two million dollars for this."

*   *   *

Three weeks later the world caught fire with the news that scientists had received a message from an extraterrestrial intelligence. Humankind had apparently entered a new age.

Mark answered the telephone at six in the morning, before he had a chance to start thinking about what he had to do that day.

"Mark, this is an emergency!" shouted SETI's new president, Bob Schmidt. Mark grimaced and pulled the phone from his ear. "Put Margaret on!" continued the voice.

Mark dutifully handed the phone to Margaret, telling her, "It's an emergency."

He watched Margaret go from leaning on one elbow, to sitting up, to jumping up and pacing the room. "You're not serious . . . I don't believe it . . . You're kidding me . . ."

"What? What?" said Mark.

"You're putting me on," she continued. "Well, that's got to be an interstellar beacon then. All right, all right, I'll be right down." She hung up and dashed for her closet. "You'll have to get Michelle to day care when it opens."

"What's going on?"

"You won't believe it."

"Did you say an interstellar beacon?"

Margaret was jumping up and down and setting a personal record for getting dressed. "Epsilon Eridani is sending out huge flashes in Morse Code."

"Morse code!"

"Some kid saw it last night. Everyone in the world knows about it but us."

"A kid could see it?"

"It's already in the morning newspapers in Pittsburgh and Chicago."

"Have they deciphered it?"

"It's in Morse Code!"

Mark followed her around the room. "You mean it's an intentional signal, just to us?"

"They've obviously been listening to us first."

"What's the message?"

"1.415 gigahertz."

"That's the message?"

"Yes. It's a beacon, to get our attention and tell us what frequency to tune into there."

"Holy—"

"And we've got a strong microwave signal at that wavelength. They're trying to decipher it now. The Institute is already filled with reporters. Get Michelle to day care and the others to school." She ran out the bedroom door.

"Yes, memsahib." Mark knew he couldn't reason with Margaret at a time like this, but he also knew that he was needed down at the Institute now even more than she was. If word was already getting around the country that some kid found the signal before SETI did, and if this frequency was so easy to pick up that others were picking it up now too, then it was crucial that SETI maintain its image as the official leader in the reception and in the decoding. The media had to know that SETI was the one authority, the one trusted source of information on this thing. All future funding depended on it.

Mark tried calling the Institute, but all the lines were busy. He didn't trust the new SETI president to handle this by himself. He called his neighbors, whom he had talked to only once in the last year, and offered them $200 to get his kids to school and day care while he took care of an emergency.

When Mark pulled up to the SETI Institute's small headquarters, the parking lot and street were already jammed with news vans. As he had feared, the Institute people were all preoccupied with the signal, leaving just Margaret and a technician to represent SETI at a crowded press briefing in the conference room. Mark called the agency that he had chosen to put SETI's publicity plan into action. Without funding or approval, he had arranged for an all-out media blitz, should such a signal ever be confirmed. He had reasoned that if the event ever actually occurred, there would be plenty of new funds to work with—they could probably create a whole new publicity department.

Mark told his contact at the agency that he would call him right back with a statement and that they should get ready to release it immediately— worldwide. He squeezed his way into the conference room and started writing a statement from what he could piece together.

Reporters shot question after question at Margaret and the technician. The room was getting hot from the lights and the bodies, but Margaret remained cool. She seemed to be made for this role. "How do we know that Gracie's message isn't some kind of elaborate Caltech prank?" one reporter wanted to know.

"If the signal is a prank," said Margaret, "then the prank is originating from outside our solar system—it can't be coming from Caltech or anywhere else on earth. That much we know. Radio telescopes in Arizona, Massachusetts, Puerto Rico, and Buenos Aires are all confirming that the signal is indeed coming from Epsilon Eridani, eleven light years away."

"Why did a seven-year-old girl in Pittsburgh discover this signal before you, with all your high resolution digital receivers scanning tens of millions of radio channels?"

Mark winced, but reminded himself that these were merely local reporters. The national news people would come later, and there would be time for Margaret and the others to be briefed on how to bring out SETI's central role in all this.

But Margaret handled the question quite well, Mark thought: "The Mount Wilson observatory has an educational program that makes its telescopes available by computer link to schools across the country. The little girl happened to be the first one to look at Epsilon Eridani with enough magnification since the beacon started. But her teacher knew to come to us to find and decode the real message. SETI has been preparing for twenty years for every possible contingency, including this one."

"Then you were expecting this kind of a signal?"

"Well, we've been preparing ourselves to expect the unexpected, but we certainly didn't expect a Morse code beacon. And we didn't expect something beamed right at us. We were expecting a weak, incidental signal, something used for another civilization's own purposes and leaked to us, something that might be beyond our ability to decipher for a long time."

"What can you tell us so far about the radio signal?"

The technician answered: "It appears to be some kind of detailed image. We were hoping the picture would decode itself, once we determined its frame size. But it doesn't seem to be using any kind of binary system. The last I was in there, it was beginning to make more sense as a scanned image, like in television."

Another reporter who had just squeezed his way into the back of the room shouted: "We just got word from an overseas bureau that Chinese scientists are getting an image by just plugging their regular television sets into their microwave receivers. Is that possible?"

While the technician tried to answer the question, Margaret excused herself and headed for the radio room. Mark followed her.

"They're getting an image on the PAL system," Margaret told Bob Schmidt, SETI's president.

A technician turned from a row of monitors. "I can rig that up."

"What's going on?" Mark asked.

"PAL is the standard television system used in China," said Margaret, "and in most of the world. In the West we use NTSC—"

"Which we already tried," said Bob. Technicians rushed back and forth with television components.

"How long will this take?" asked Mark.

"It's a simple format conversion," said a technician. "Maybe five minutes."

"You've got to make it faster," said Mark, hurrying back out the door.

As Mark started past the conference room, he noticed that all the news people were now surrounding the one reporter who had news on the image the Chinese scientists were receiving.

"This isn't confirmed yet," said the reporter, "but our source says it's some kind of script. It looks a lot like Arabic or Hebrew or something."

Someone grabbed Mark by the elbow to stop him from rushing down the hall, and he turned to see a harried-looking secretary. "Mark, it's the Sultan of Brunei for you on line two."

Mark took the call in the president's office. "Your Majesty, you've heard the news."

"Yes, Mr. Rizley. It would appear that you made your find without yet doubling the spectrometer. Perhaps that's a million dollars you won't need."

He's playing games with me again, thought Mark. "As a matter of fact, we've never been in *greater* need. This is no longer about increasing our chances of finding something. This is about detecting and decoding what we know is there. We now have reason to build truly high-powered equipment in order to detect all the other signals coming from the same spot. Any advanced civilization is going to have lots of signals—radio, television, probably in color and in three dimensions." Mark paused to compose his crucial pitch: "Your Excellency, your gift could be the one that will give us our first picture of an extraterrestrial being. We're going to need tens of millions now. Are you going to be a part of this? Can we count on your help?"

Mark waited through a pause that lasted almost long enough to make him check to see that the line hadn't gone dead. Finally the Sultan replied: "Yes, Mr. Rizley, I think you can count on me for more support. Now tell me—what is the message from this signal?"

"Well, it's—uh, it's a written message—I think they're just about to announce it . . ."

"My friends in Saudi Arabia tell me it's in Hebrew. Is that true?"

"Yes, as a matter of fact, it . . . uh, they have the message now, I understand, but the reporters have been keeping me from getting back to them. Would you like me to call you right back with the message?"

"By all means. But please don't keep me waiting long. It's already past midnight here and I won't sleep until I know."

Mark ran back to the radio room and poked his head in the door. There was the message on the monitor: two lines of some kind of oriental-looking script, crystal-clear. The scientists puzzled over it.

"Don't any of you know Hebrew?" Mark blurted out. The scientists and technicians looked at one another. "Rosenblum, weren't you paying attention in Hebrew school?"

"I never went to Hebrew school," said Rosenblum. He looked closer at the screen. "I thought it looked familiar."

Mark remembered that there was a synagogue just three blocks away, and he was on his way out the door again. The secretary grabbed him once more as he sprinted around a corner.

"David Packard of Hewlett-Packard is on line 2 and Paul Allen of Microsoft is on line 3 for you."

"Tell them I'm out but I'll be right back." The thought that these contributors could only be calling to offer more support gave Mark added energy as he ran down the street. He went over his list of potential donors in his mind. He had fifty people he could practically count on for a million or more after this announcement. Probably twice as many foundations. So many were so close to contributing as it was; this news would surely put them over the edge. He leaped over a hedge and began calculating the average gift he could expect from them and his percentage. It's official, thought Mark, I can now count myself among the *nouveaux riches*. Margaret and I will be celebrating tonight.

Mark ran into the rabbi's office with a policeman at his heels. The officer wanted to know what he was running from. The policeman ended up giving Mark and the rabbi a fast ride in his squad car back to the SETI Institute.

A holy hush swept over the radio room as the rabbi peered into the screen. He looked back at Mark. "This is your message from outer space?" asked the rabbi.

"Yes. What's it say?"

"It's the first verse of the Torah. It says, 'In the beginning God created the heavens and the earth.'"

*   *   *

Had it not been for Mark Rizley, Gracie's message would have stayed with the group in the radio room for hours—and others would have announced it first. The SETI people insisted on confirming the message with more radio receivers. Bob Schmidt didn't want the Institute to look foolish in case there eventually turned out to be some other explanation. The scientists discussed various and devious means by which a sophisticated prankster could have meddled with their equipment, and for a while everyone suspected everyone else of being in on a great hoax.

A SETI Institute technician inspects Project Phoenix's Targeted Search System inside the Mobile Research Facility. *Courtesy of SETI Institute*

But at the height of the paranoia, Mark slipped out of the room and quietly placed his call to the agency. In fifteen minutes' time, the news had reached every major newspaper and broadcasting network in the world.

Mark was pleased to see that the word "SETI" got into the headlines almost half the time, and the SETI Institute figured prominently as the leader in the discovery in about ninety-five percent of the stories. History had been made.

Rabbis, pastors, and priests suddenly found themselves guests on radio and television talk shows, trying to explain the significance of Gracie's message. Perhaps the Creator Himself had decided that things had gone far enough on earth and that people needed a special reminder to bring them back to basics in morality and belief. One obscure, Bible-believing apologist by the name of Fred Heeren claimed that all the world would now have to believe in the God of the Bible.

The world, however, was more cautious—especially the scientists. Mount Wilson's famous astronomer, Robert Jastrow, was quoted as saying that the signal was consistent with the idea of a Creator. Nobel-prize winning physicist Arno Penzias told reporters that he could only say that the signal was not *inconsistent* with the idea of a Creator. And at the University of Cambridge, Stephen Hawking said that, for all we know, the signal had simply been broadcasting forever. So long as the signal had a beginning, we could suppose it required an intelligence to initiate it. But if the signal is simply playing in some sort of endless loop, what place then for a Creator?

Most scientists believed that the signal must have had a beginning, but they explained it as a very great coincidence. No matter how great the odds seemed to be against the chance that natural radiation could ever produce such a perfectly designed message, this still appeared to be the only scientific explanation. There might be many universes, they reasoned—perhaps an infinite number of them—in which case *everything* must eventually happen in one of them. One of these universes would have to produce such a signal from natural microwave emissions.

But once all the world's microwave receivers were trained on Epsilon Eridani, the signal changed. It continued on through Genesis, displaying the next sentence and then the next, making its way through the Hebrew Bible at a pace of about a page of Bible text each day. Again, the cosmologists explained that, given an infinite number of universes, one of them was bound to form page-by-page pictures of the Book of Genesis from this star's natural microwave radiation.

The SETI people felt torn: they wanted to defend the legitimacy and the importance of the extraterrestrial intelligence they had discovered, but no one could bring himself to face the possibility that God Himself was that extraterrestrial intelligence.

"My own best guess is that this *is* a hoax," Bob Schmidt told his board of directors at their first meeting after the discovery. "But, as we have confirmed, the hoax is not being perpetrated from within our solar system. Some advanced civilization on a planet orbiting Epsilon Eridani must have picked up signals containing our Bible and decided to beam them back to us, just to make us think it was coming from God."

"Why would they do that?" asked SETI's treasurer.

"Perhaps they have designs on us or our planet," said the president, "and they just want to make us more religious so we don't blow ourselves up before they can get here."

The directors considered the implications. "You mean," said the vice president, "that maybe this alien civilization wants to make us their slaves . . . or eat us?"

"Or cultivate our planet before we spoil it with atomic radiation," said the president.

The recording secretary suddenly stopped scribbling in order to consider the implications too.

"I like it," said Mark. "This might give the whole thing a new angle. Mark ignored the raised eyebrows and went on: "I mean, we're in an emergency situation here. There's no suspense in it anymore. Donors are telling me, 'We don't need to pay a million dollars to read the Bible.' Something like this would really raise people's awareness."

In the end, the board decided to avoid creating a panic. Mark continued to worry about the dwindling enthusiasm among contributors, and rightly so. Firm commitments became tentative and tentative offers disappeared. After all, Gracie's message was now being received by many microwave receivers and television sets all over the world without the Institute's help. The Associated Press, however, continued to pick up its daily message from the Hebrew translators hired by the Institute, and people all over the U.S. read the message in a daily newspaper column that usually ran right next to the lottery numbers.

Critical scholars had a field day with the text. They noted that Gracie's message followed something close to the Hebrew Masoretic wording. And by the time the daily broadcasts reached the book of Hosea, Jewish scholars pointed out that Gracie's Bible books followed the Jewish order, not the order in the Christian Bible.

Difficult passages were sometimes settled on the basis of the extraterrestrial message, but more often, disputed passages remained disputed, since translators still could not agree on the fine points of certain Hebrew meanings. Major teachings, which were not in dispute, interested the scholars less and were seldom considered or discussed.

Muslims were upset that the signal was not the Koran and Hindus that it was not the Upanishads—and many Christians were upset that it was not the authorized King James version.

But for the average, open-minded Earthling who was just interested in getting at the truth, Gracie's message provided the impetus for the ultimate Bible study. People from Singapore to Stockholm opened up their newspapers every day, read the message and discussed it as if it were a new bestseller. U.S. media people, educators, statesmen and scientists began to find out for the first time what the Bible actually had to say. Even Margaret Rizley began to take a new interest in the God of the Bible.

*   *   *

A year-and-a-half later, Gracie's message reached Psalm 19. Margaret was reading it when Mark's morning alarm sounded. He switched it off and lay back to brace himself for the day ahead. It was all the more important that he succeed in finding more outside grants soon, now that SETI could only afford to employ Margaret part-time.

"Thinking about everything you have to do today?" asked Margaret. Mark looked up to see her reading next to their bedroom window.

"Yeah, I need to finish the proposal for the laser gene mapping, which has little chance of getting funded, and then I need to badger all those millionaires about keeping their commitments to SETI, which they probably won't."

"You can count on the Sultan, anyway."

"He's about the only one," said Mark, getting up. "And I don't know if he'll stay committed if you folks can't find some trace of another signal from Epsilon Eridani pretty soon."

"The Sultan sounded pretty committed to SETI the last time we talked to him," she said. She followed Mark to the bathroom and watched him shave. "Sometimes I wonder if that's the *only* kind of research he's helping us with." And then she deepened her voice and put on her best British accent to do her impression of the Sultan: "What do each of you really want, hmm? My money, of course, but why? You, Mrs. Rizley, you want my money for research. You would be content to do research forever, as long as it took you only where you wanted to go."

"You sound more like Sean Connery."

"And you, Mr. Rizley . . . do you actually think you will find contentment with what money can buy?"

"You'd be the last person to know, Majesty," said Mark. "You don't have to get up in the morning and face mortgage payments, car payments, credit card payments, while trying to save up for your kids' college and for retirement."

"Maybe he's the *best* person to know," Margaret thought out loud, and then she resumed her impression: "But Mr. Rizley, why spend money on what is not bread, and your labor on what does not satisfy?"

Mark stopped shaving to look at Margaret. "Did the Sultan say that?"

"No, that last part was from Gracie's message. I've been reading ahead." Margaret walked away and reappeared a few moments later with the newspaper. "You know," she said, "if you really read this, you can't help wondering if this might actually be coming directly from . . . you know."

"God?"

"Well, yeah. What if this signal is His way of telling us, "Here's the extraterrestrial intelligence you've been looking for"?

Mark resumed shaving. "Don't flatter yourself that it's a personal message from Him just to you or to SETI. I mean, the whole world's getting it."

"That brings up an interesting question," said Margaret, "in light of what I was reading today." She found her place in the newspaper column and asked, "Do you think God is fair?"

"Well, I never thought of Him as being *dark*."

Margaret ignored him and continued: "I used to think the God of the Bible was unfair if He expected people to believe in Him on the basis of the Bible when not everyone *has* the Bible."

"And they still don't," said Mark, drying off his face. "Some of those Muslim countries are banning Gracie's message. So if the Bible's God is God, then He's not fair to everyone and the whole Judeo-Christian message falls apart."

"But what if He's been telling us about Himself all along, just by His perfect handiwork up there, compared to the mess we've made of the world down here? I mean, everyone who knows cosmology knows this universe is set up within very carefully chosen parameters that make stars and galaxies and life possible—

"Everyone doesn't know cosmology."

"But *everyone* can look up and say, 'Hey, all this couldn't have come from nothing.' That's the point of today's message—it's for *everyone*." Margaret read from the morning paper:

> The heavens declare the glory of God;
>   the skies proclaim the work of His hands.
> Day after day they pour forth speech;
>   night after night they display knowledge.
> There is no speech or language
>   where their voice is not heard.
> Their voice goes out into all the earth,
>   their words to the ends of the world.

"So?" said Mark.

"So maybe God is fair."

"Do you think He's been fair to us?"

"I guess that depends on what's important," said Margaret. "Maybe the most important thing is that He gets our attention."

Mark headed into the shower. "How come every time I talk to you I feel like I'm in the middle of an Ingmar Bergman movie?"

Mark soon forgot most of that morning's discussion, but a similar passage got his attention a few months later. While driving home from work and hunting for a radio station without a commercial, he heard a portion of Gracie's message:

> Lift your eyes and look to the heavens:
> Who created all these?

Mark thought back to the night long ago when he had looked up at the stars and asked God for a sign that He was there. What if the stars themselves had been the sign, the signature of a great Creator and Sustainer, the signal proving that humans were not alone in the universe? Perhaps this perfectly designed universe itself was the greatest sign anyone could ask for. Perhaps all that cosmologists had learned about the need for a beginning of time and space was pointing to the need for a Creator *outside* of time and space.

In that case, Mark realized, this universe would not be all there is. There would be a timeless realm, a spiritual world that the Spirit of God and his own spirit could also experience. But that would make most of what Mark had worked for all his life look very small. The most important concerns in his life would actually be the least important, and the least important would be the most important. Mark tried to compare the amount of time he spent on self-interests with the time he spent developing spiritual interests or godly traits or whatever God expected of him. The required change seemed too enormous to contemplate, and he succeeded in putting it all out of his mind.

* * *

After two-and-a-half years of steady broadcasting, the microwave emissions from Epsilon Eridani abruptly stopped. The Hebrew Bible had been completely transmitted. Many people considered converting to Judaism. Mark Rizley joked: "What do you know? Sammy Davis, Jr. was right."

According to "International SETI Post-detection Protocol," the nations of the earth were supposed to decide together whether and how to reply. Bringing the nations of the Earth to an agreement about the reply was not an easy matter, however. A SETI representative returned from each United Nations meeting without success. If Gracie's message had been anything else, a reply of general goodwill would have been relatively simple, but what were the nations of earth to say to *this*?

During the first year of the transmission, U.S. school children had been urged to think about how they would reply in their science and English classes. During the second year, however, Congress passed laws prohibiting children from talking about it during class time, since atheists and people of non-biblical faiths might be offended.

Certain Arab nations, which had been attempting to suppress the message in their countries, were especially difficult to please in composing a response. But a month after the transmission ended, after some last-minute haggling, the United Nations finally succeeded in producing a two-page reply, the gist of which was, "What does this mean? What do you expect us to do with this message?"

The Arecibo observatory in Puerto Rico beamed out the message, in Hebrew, to Epsilon Eridani, and then the world prepared to wait through the twenty-two year delay until they might receive an answer. Since Epsilon Eridani was eleven light-years away, it would take eleven years for this signal to make the trip there and then another eleven years for the return reply. Older folks, many of whom seemed most anxious to hear the return message, were unfortunately the least likely to live long enough receive it.

What *was* Gracie's message supposed to do for us? people wondered. Just tell us we're all sinners, incapable of keeping our end of a covenant with God? What about the promises it contained of once-for-all forgiveness and of a future, everlasting covenant? What does God want from us *now*?

In the very same hour that the antenna in Puerto Rico began to transmit Earth's query, the world's microwave dishes started to receive a reply from Epsilon Eridani, this time in ancient Greek: The New Testament.

"The beginning of the good news about Jesus Christ, the Son of God," said the first message. It was the first sentence of the gospel of Mark. The rest of the New Testament followed. Again the textual scholars had a field day with the order of the Bible books and the textual details. Christians of every denomination looked for textual support for their various positions.

Agnostics and people of non-biblical faiths claimed that this latest signal was proof positive that Gracie's message had to be a grand hoax. It would be impossible for Epsilon Eridani to receive Earth's message and send it back before twenty-two years had elapsed. People argued about whether even God was capable of sending a message faster than the speed of light. A consensus grew among scientists that the signal had to have its source on earth. If it was actually coming from Epsilon Eridani, then perhaps someone had found a way to beam it there from Earth so that it would be reflected back. Perhaps the timing of the reply had been a coincidence.

But then microwave receivers pointed at other stars started picking up the message too. In fact, radio astronomers soon realized, Gracie's message was now coming at them from *all* points of the sky. Like the microwave background radiation, which since 1965 has yielded evidence of the universe's creation event from all points of the sky, this message was difficult to avoid.

Some scientists continued to insist that, given an infinite number of universes, there would have to be one in which all the stars would naturally radiate just the right series of pulses in order to produce the Bible as a scanned television image.

By this time, however, most average folks around the world accepted the signal as a genuine message from God. For some, Gracie's message became better known as "the Message of Grace." But as the broadcast progressed, many thought of it as a message for *other* people. It was for the bad people, not them; it was for the people who needed a good dose of religion to straighten them out. As it turned out then, the only ones who turned to God in faith were those who probably would have done so anyway. Life on earth continued almost as usual.

Reading the extraterrestrial column every day, Margaret Rizley was among those who began taking Gracie's message to heart. Mark decided that a little religion might be good for her. God knows she needs help controlling her temper with the kids, he thought. From Mark's viewpoint, Margaret's mood swings could be pretty wild, and at her worst she subjected the children to what could probably be described as verbal abuse in a court of law.

One Saturday morning, when Mark woke up without Margaret beside him, he groaned with the realization of what he had to do that day. Margaret had gone to a weekend astronomy conference at Berkeley, leaving him to take care of the three children by himself. By the end of that weekend, Mark had a new understanding of his own dark side. In fact, by Saturday evening, he had already verbally abused each child many times, by anyone's definition. He had been amply provoked, of course. The three-year-old had spilled her entire plate of food twice, the eight-year-old had interrupted him incessantly during important phone calls, and the ten-year-old had pinched his brother's fingers in the door, requiring a trip to the emergency room. He never did find out which one had destroyed a file while playing with his computer, since all swore innocence. Mark wondered if genetic engineers would one day identify the "bratty gene" and eliminate it from the human race.

When Margaret returned Sunday night, she asked how everything went.

"Oh, okay. They were a little rambunctious after they went down tonight. So I had to had to take away their dessert privileges for tomorrow." He neglected to tell her that there had been so much shouting and wailing a few minutes before that the police had stopped by to see if everything was all right.

Margaret cleared a place for herself on the living room couch and took stock of her house, which was cluttered with toys, clothes, and crayons. "I bet this was a good weekend for research, huh?"

Mark seldom acknowledged Margaret's running joke about life's research project. "How'd the conference go?" he asked.

"It turned into a big argument about Gracie's message." She cleared a place for Mark to sit down next to her. "You know," she said, "sometimes I think all this research is just about getting us to ask the right questions: 'Is there any justice in the universe? How could all this mess on earth ever be straightened out? How would a perfect judge judge me?'"

Mark sat down beside her. "I guess all anyone has to do to start a religion is make up good-sounding answers to all those universal questions."

Margaret reached in her purse and pulled out a pocket-sized edition of Gracie's message. "But this is the one message that comes to us from ancient times that could possibly come from the real God."

"You mean as far as getting the cosmology right."

"Absolutely," said Margaret. "All the other ancient religions said that the universe had no beginning, or that there was some kind of watery substance that came first and that the gods came later out of that. If God has ever spoken, this is it."

"You really think the Creator of the whole universe would take the time to speak to us little earthlings?"

"I think God is either a sadist who created us with this hopeless need to know Him, or He's everything Gracie's message says. I think living is meant to teach us that this is what we need. Not just a far-off, Creator-God, but a Counselor, a Father . . . a Savior." She looked at Mark, anxious for his reactions.

"I wouldn't want to try to talk you out of it," said Mark. "Some people need to have their sins forgiven. Some people need an invisible friend to talk to. And if God really does talk to some people, well then God bless 'em . . . but He doesn't talk to me."

Margaret looked down at her Bible and asked, "What about this?"

"Is that enough for you?"

"Yes, it is. Maybe if you'd really read it—"

"I think I need to know that God cares about us more than just to create a nice world, watch it get messy, and then send us down a rule book."

"What He sent down wasn't just a book," said Margaret. "It was a person."

Margaret went on to explain what she had been studying about the

Old Testament prophecies of a Messiah. They had foretold that a "ruler from eternity" would be born in Bethlehem, that He would be rejected by His people and killed just before the temple was destroyed. Mark's mind wandered while she recited prophetic details that had impressed her, but then he engaged it again when she read from Isaiah 53:

> *But He was pierced for our transgressions,*
> *He was crushed for our iniquities;*
> *the punishment that brought us peace was upon Him,*
> *and by His wounds we are healed.*
> *We all, like sheep, have gone astray,*
> *each of us has turned to his own way;*
> *and the LORD has laid on Him the iniquity of us all.*

Here we go, thought Mark. Iniquity, transgressions, Jesus died for my sins—it all had that old-fashioned, fanatical sound to it. If Margaret took this too far and became vocal about it with her colleagues, she might make a real fool of herself and ruin her scientific career.

"Margaret," he interrupted, searching for a way to explain the difference between redneck extremists and people like themselves, "I want you to just consider the possibility that this may not be for you. You're one of the most moral people I know. And I'm one of the other most moral people I know."

It was perfectly true, he knew. As people go, Mark was more moral than most. In fact, when Mark thought about it, he was *better* than moral—he was *balanced*. Not too wicked, not too holier-than-thou. To get any more religious than he already was would spoil it. He remembered for a moment how the weekend had revealed his dark side, but surely his little vices were paltry compared with the evils that he heard about on the nightly news.

"Well, then maybe we're about perfect," said Margaret. "Maybe neither of us has any problems with pride or hate, or with envying other people, or with adulterous thoughts."

Mark wondered how much Margaret could guess about his fantasy life, but he said, "Do you hear yourself? You're sounding like, like some kind of street preacher."

Margaret became quiet for a while. Mark wondered what he would have to do to satisfy her on this matter. It was hard for him to conceive of himself as becoming any more moral than he already was, but if he were to actually go to the bother of trying, he decided he would at least like to be absolutely sure about the truth of all this. He thought about the time he had waited in vain for a sign from God among the stars.

"Tell you what," said Mark. "If you ask God to please come and appear before us right now, and He appears, then I'll admit I'm a rotten sinner and dedicate the rest of my sorry life to Him."

"You've said that before," said Margaret, "and I've thought about it."

"So let's try it—right now." Mark stood up and looked up at the ceiling. "God, if you really get involved in individual cases, how about this one? I'll do whatever you want me to do if you just come and appear at our front door in all your glory." He walked to the front door and opened it.

"Looking left, looking right, looking up. Anybody there?" He caught a glimpse of the Big Dipper and then paused a moment to let his eyes get better adjusted.

Margaret joined him at the open door. "If God were to personally appear before you, in any form that was spectacular enough to convince you that it was really Him, then you wouldn't have any real *choice* about following Him—you'd know you *had* to. And when there's no choice, there's no opportunity for love. But that's what He wants. He doesn't want to *make* you follow Him."

Mark looked for holes in the argument but found none. He closed the door.

"There are worse things an extraterrestrial intelligence could want from us," said Margaret, "besides our love and our trust."

"Love and trust have to be based on something tangible, something you can get involved with. It has to go both ways. Other than the fact that the universe's design and beginning and all that seems to require a God, we have no way of knowing He's really involved or cares."

"But that's what Gracie's message is all about," said Margaret. "He's gone to the greatest *possible* length to show that He cares, that He's involved: He's become one of us. And He's done the greatest thing one person could ever do for another: He's given up His life for us. He's given us the best possible reason to return His love. And that's all He wants. To put it in your terms, could you ever think of a better deal? What's the downside risk?"

But Mark was already calculating the risks. All the personal freedoms he might have to give up. The risk of looking like a fanatic before his friends. And there was always the chance that there was still some other explanation for the origin of the universe, as yet undiscovered. Then he would be risking looking like a fool for nothing.

Of course, if it turned out that it actually *was* true . . . . Well, there would be more time to make up his mind later.

*   *   *

Shortly after Gracie's message stopped broadcasting, the story passed from the newscasts to the history books. Some continued to believe; others wrote about it as an elaborate hoax or coincidence.

The United Nations never did come up with a reply to the second signal, but millions of individuals all over the world did. They beamed up their return responses via prayer on a minute-by-minute basis. They were the only ones who ever achieved two-way communication with an extraterrestrial intelligence. And most of them claimed that it was an intelligence that could indeed solve all the problems of the world—starting with each individual human heart.

Mark Rizley occasionally sent up signals too—whenever a crucial business deal loomed before him or illness struck. As he approached retirement, he prayed for the day when he would be able to finally relax and enjoy all the comforts for which he had worked so hard.

More often as he got older, tragedies came into Mark's life. Sometimes the death of a loved one affected him so deeply that it drove him back to his window at three o'clock in the morning. There he waited for a message from God—a special miracle just for him alone—to tell him what he was supposed to be getting out of life's research project and what God wanted out of him.

Soon even the cries for a quick fix in emergencies ceased. Mark's heart had become as cold as outer space, as silent as the moon.

*   *   *

All interested parties looking in finally saw the matter clearly: even a sign from heaven had not been enough for Mark Rizley. These affairs of human hearts, after all, were things that even the messengers from other realms desired to look into. To them it was another story of unrequited love. To those members of Mark's race who had been waiting to welcome him into new realms, Mark was a great disappointment; though none—not even Margaret—would have wanted him there without his heart. To most, the idea of sending Mark a message from space had seemed a great extravagance. But then, they weren't running the universe. And each of them knew that their Father had been just as extravagant in their cases.

*   *   *

At the age of 97, Mark was a living testimony to the marvels of modern medical science. Most of the rest of his family had not fared so well, and Mark could never remember which of his children, if any, were still alive. While pondering this question one bright morning, Mark gradually became aware that a young man was standing in front of him, trying to tell him something.

"Great Gramps, remember I said you were having some special visitors today?"

"Now what?" croaked Mark. "What do I have to do today?"

"They're from the television station and they want to talk to you about your work on SETI."

"About my work? My work never stops."

"They're right outside. Is it okay if I let them come in now?"

"Who? Who's that?"

"The people from the television station."

"I don't know. I've got a lot to do." A nurse's aide took his breakfast tray away. "Hey, I wasn't finished!"

"It looks done to me."

The young man rose and went to the door. He talked lowly with the reporter and the cameraman and led them in.

"We'll have to make this fast," said Mark. "I've got a million things going on at once here."

The reporter sat down across from Mark, and after introducing himself, he said, "Mr. Rizley, scientists now have several good suggestions for how it might be possible to duplicate the sending of microwave signals from the stars to earth. Do you believe Gracie's message came from earth, or from an extraterrestrial source?"

Mark tried to look beyond the bright light to find the young man again. "Listen, Jason, if I don't get back to work, you won't have money for college."

The young man leaned forward and told the reporter: "Jason was my grandfather."

"Can he understand our questions?" asked the reporter.

"Sometimes."

The reporter turned back to Mark for another try. He talked slowly. "Mr. Rizley, you're the last person we can talk to who was involved with the old SETI team. There are rumors about a conspiracy among the SETI people to perpetrate a hoax."

"Margaret gave her life to SETI, you know," said Mark.

"Did Margaret help to send that message?" asked the reporter.

"No," said Mark, "she *got* the message."

"Did you get the message too?"

Mark scanned the embroidered plaques that covered his walls. There was a verse about remembering your Creator before the silver cord is severed, another about buying things without money. None of it made much sense.

"Margaret made those," said Mark. "I don't have time to read all that. I'm too busy making a living here to stop and worry about all that. Who's going to pay the rent if I don't?"

The cameraman aimed his camera at the hand-stitched verse that hung over Mark's head:

> Come, all you who are thirsty, come to the waters;
> and you who have no money, come, buy and eat!
> Come, buy wine and milk without money and without cost.
> Why spend money on what is not bread,
> and your labor on what does not satisfy?
>
> *Isaiah 55:1-2*

"Mr. Rizley, do you think we're alone in the universe?"

"I've got bigger things to think about. See if you can get some extraterrestrial intelligence to pay the rent. Otherwise, I don't need any extra intelligence." Mark noticed that the cameraman continued to take shots of the wall plaques. "My glasses don't work right. Michelle reads those to me. Is Michelle here?"

"That was my great aunt," whispered the young man.

Overhearing this, Mark had a lucid moment when he recalled that his daughter had died a long time ago. "Well if you're going to start talking behind my back," he growled, "why don't you all get back to your business and let me get back to mine? Where are my papers now?"

Mark's great-grandson handed him a stack of papers, and Mark started shuffling them dutifully.

Part III

God and the Origin of Everything

Carl: Uh-oh. Now you're going to scare people away. Now you're going to start talking about the C word.

Fred: You mean cosmology?

Carl: Shhhh. Let's face it. Cosmology is not something one talks about at the beauty parlor while under the hair dryer. The redshifting of light from galaxies seldom seems to come up at neighborhood block parties.

Fred: Actually, before we get into the science of it, I thought we'd start with some commonsense reasoning about God.

Carl: Maybe if we could dig up some kind of scandal about Stephen Hawking or Carl Sagan, then we could get people to talk about it under the hair dryer. Hey, how's this for a title? *The Latest Dirt on the Origin of the Universe.*

Fred: What I want to start with is: How far can common sense get us? Can logic tell us anything about whether there is a God?

Carl: Common sense? Logic? You lost me already.

Fred: And if there is a God, can logic tell us anything about what kind of a God?

The Andromeda Galaxy, M31. This farthest object we can see with the naked eye is the nearest galaxy to us (2.3 billion light-years away). With about 7 spiral arms, it resembles our own Milky Way. —*Courtesy of Lick Observatory, University of California*

CHAPTER 4

Is the Bible's God the Best Explanation?

Why is there something rather than nothing?
—Lucretius

A Skeptic's Questions:

Didn't the cosmological argument lose popularity a long time ago among philosophers? Isn't it possible that the universe is eternal—that it just *is*—and so there's no need for a Creator? And even if there *is* a Creator, isn't it just as likely that God is impersonal, that God is just a force, an organizing principle—that God is simply *nature*?

A Bible Believer's Response:

As I can show in more detail below, the universe could not have caused itself. All that we observe around us is proof of an uncaused Cause. Nothing in cosmology or philosophy has changed this fact. Philosophers may have moved on to concentrate on other problems (leaving the origin of the universe to the scientists), but it's not because they've solved this one.

And as we will see below, logic leads us to believe that this First Cause must be separate from what it created, transcending it—that it must be eternal, spiritual, all-powerful, all-knowing, purposing tremendous undertakings on behalf of human beings, personal—in short, that this First Cause is more perfectly explained by the God of the Bible than by anything else.

QUESTION: Is the God of the Bible
the most logical explanation for the origin of the universe?

Logic Demands a Cause for Every Effect.

This is not rocket science. This is common sense, and no one has ever observed an exception. Even Julie Andrews sings about it: "Nothing comes from nothing; nothing ever could."

That every effect must have a cause is a self-evident truth, not only for those who have been trained in logic, but for thinking people everywhere. The cosmological argument for God is founded upon the old Latin axiom, *Ex nihilo nihil fit:* From nothing, nothing comes.

The Universe Is an Effect Which Demands a Very Great Cause.

No big bang theory or oscillating universe theory or static universe theory (all to be discussed later) has gotten around the need for an ultimate cause. A series of causes cannot be infinite. There must have been a first cause, which itself is uncaused.

The existentialist philosopher Martin Heidegger aptly posed the question common to every atheist who has ever stopped to consider how the universe came to be: "Why is there any Being at all—why not far rather Nothing?"[1] "How extraordinary that anything should exist," echoes philosopher Ludwig Wittgenstein.[2]

Isaiah directed us to lift our eyes to the heavens and ask the universal question, "Who created all these?" (Isaiah 40:26). David wrote, "The heavens declare the glory of God; the skies proclaim the work of his hands. Day after day they pour forth speech; night after night they display knowledge. There is no speech or language where their voice is not heard" (Psalm 19:1-3).

And Paul said, ". . . what may be known about God is plain to them, because God has made it plain to them. For since the creation of the world God's invisible qualities—His eternal power and divine nature—have been clearly seen, being understood from what has been made, so that men are without excuse" (Romans 1:19–20).

The Bible says that when we look into the heavens, we find a self-evident truth there, as obvious as if we could hear it in words (Psalm 19:4), available to people of every language, to every part of the earth, and we're without excuse if we don't believe it.

This self-evident truth is the simple, rational deduction that all we see is an *effect* which demands a very great, supernatural Cause. The sun and the stars, the moon and this earth could not have come from nothing. That's irrational—not just to the Western mind, but to the *human* mind.

Every phenomenon in the universe can be explained in terms of something else that caused it. But when the phenomenon in question is the existence of the universe itself, there is nothing in the universe to explain it. No *natural* explanation.

Even if a scientist feels he can explain the evolution of life, even if he can explain the evolution of the universe from the tiniest fraction of a second after the big bang, science has no natural explanation for how matter and energy could have emerged from nothing before that—and then continued for many ages right up until the present (as opposed to the appearance and disappearance of quantum particles, to be explained in the next chapter). After thousands of years of thought, humankind is limited to the same explanation it has had from the beginning: a supernatural explanation. Something outside the universe, outside nature. Something that fits the biblical concept of God.

Skeptics may at first counter, "Then who created God? Your argument begs the question." But if the original question is, "What caused the universe?" then to answer simply that the universe *has* no cause clearly begs the question. But when we answer, "A limitless Being from outside of time and space," we are not only answering the question more directly than the atheist, but we're giving the *necessary* answer—because nothing that is already a part of the universe could have created it. The fact that we time-bound creatures can't picture how the Creator can live outside of time, without a beginning, does not remove the need for Him.

Again, when we say, "Something can't come from nothing," the atheist may counter: "But you say that God came from nothing." This is not true. Bible believers do not say that God came from nothing. According to the Bible, there was never a time before God, when there was nothing (Psalm 90:2). God always existed. The atheist may say: "Well then, there's just as much reason to believe that the *universe* always existed." Again, untrue. We have no reason to believe that the universe always existed. The universe cannot explain itself. It has no reason for being in itself. The God of the Bible, however (who calls Himself *I AM*), has specifically defined Himself from ancient times as a self-existing entity. He lives eternally, depending on nothing outside of Himself.

Now that 20th-century science has supplied us with ample evidence that our universe and time itself had a *beginning* (to be discussed in Chapters 6 and 7), we know that time is not infinite; its beginning requires an explanation. Nothing that is confined to time could cause the universe. Time, by definition, is that province where cause-and-effect events happen, where every effect must have its cause. God, by the Bible's ancient definition, is not confined to time. Thus, while events in time certainly require a cause, an entity outside of time may not.

But How Do We Know That This First Cause Is Anything Like the God of the Bible?

The First Cause Must Be Independent of Its Effect.

Logic demands that the Creator must be completely independent of His creation. The First Cause must not require any of the things that depend on *it* for *their* existence. Otherwise there would still be nothing. The idea of a "Star Wars" God, a God who is a mere "Force" that is one with or part of the universe, is eliminated. God must be transcendent; that is, above and beyond the boundaries of His creation.

In Eastern religious thinking, the "Supreme Soul" or "the Infinite" is taught to be the place where one sees nothing but unity, hears nothing else, understands nothing else, like the state of deep sleep. However, such a less-than-conscious, purposeless "infinity" could never account for the tremendously purposeful, super-intelligent design evident in the universe (see Chapter 9). And such a mindless "force," if it is inseparable from the universe itself, could not have been around before (or apart from) our universe in order to create it.

As we will see in coming chapters, this is the first thing the discoveries of 20th-century cosmology suggest about the universe's greatest mystery, the greatest whodunit of all time: it was an outside job.

The First Cause Must Be Infinitely Powerful (Omnipotent).

Although he was no Bible believer, the renowned philosopher Baruch Spinoza logically deduced that the First Cause had to be not only independent of its creation, but infinite as well. The First Cause had to be unlimited, because if it were limited, it would have to be limited by some other thing (it couldn't be limited by nothing), and it wouldn't be completely independent any longer. So this entity which requires nothing else for its existence must be without limits—infinite.

Also, logic tells us that an effect cannot be greater than its cause. Thus the First Cause must be greater in power than anything in the universe; in fact, it must be greater than the sum of all the powers in the universe.

Once we accept the idea that the universe had a First Cause, we must also accept the fact that all the miracles of the Bible (from the parting of the Red Sea to the resurrection of Jesus Christ) are quite plausible and easily explained. What can be too hard for the God who created our entire universe?

The First Cause Must Be Eternal (Transcending Time).

The Creator must exist outside of time. Nothing *in* the universe can go back before the creation event, but the Creator must, if He started the process. From our perspective, He is without beginning or end. And from His perspective, outside of time, beginnings and endings are meaningless. He simply *is*. In the Hebrew Bible, God describes Himself as "I AM" (Exodus 3:14). In the New Testament, Jesus said, "Before Abraham was, I am" (John 8:58). Notice how Jesus breaks our rules of grammatical tenses in order to express, not only the divine Name, but His timelessness.

In the Exodus passage, it is interesting to note that when Moses asks God for His name, the answer succinctly implies both His timelessness and His uniquely divine, causeless existence: "I AM WHO I AM." He is the necessary One who owes His existence to nothing else. And so He instructs Moses, "This is what you are to say to the Israelites: 'I AM has sent me to you.'" "Yahweh," the Hebrew word we translate "LORD" in our Bible, sounds like and is apparently derived from the Hebrew for "I AM." This was the God proclaimed by the Bible while the other religions were worshipping a host of harvest gods, war gods, sun gods, moon gods, animal-headed gods, and fertility gods and goddesses.

No other religious writings tell of a deity who fits the picture of a time-less God as well as the Bible. Psalm 90:2 says that *before* God brought forth the earth and the universe, from everlasting to everlasting, He is God. Both 2 Timothy 1:9 and Titus 1:2 speak about what God purposed to do "before the beginning of time."

When the Bible writers say that with the Lord a day is like a thousand years, and a thousand years are like a day (2 Peter 3:8, Psalm 90:4), they proclaim a divine perspective that we only now are beginning to grasp as the relativity of time. One of the most fundamental principles of Einstein's theories is that there is no absolute time. Each observer has his own measure of time, according to his perspective.

Other contemporary religions did not view time as such a relative thing. In fact most of them had gods who were restricted to place as well as time, tribal deities who could help their people in the hills, but not on the plains, etc., according to their limited jurisdictions. As archaeologist John Romer pointed out in his PBS television series, *Testament*, the God of the Hebrews was unique among all other gods because of His ability to move through both space and time, transcending them, leading His people over vast distances from one country to another and over many generations of time.[3]

The First Cause Must Be Spiritual (Transcending Space)

Logically, we must deduce that an entity *outside* the universe is the only kind that could have created it. The First Cause may interact with the three dimensions of our space, but it must itself be beyond them, beyond the physical. Jesus emphasized that God is *Spirit*, and those who worship Him must do so in *spirit* and in truth (John 4:24).

Like archaeologist Romer, astrophysicist Sir Fred Hoyle also recognizes "the Judaeo-Christian idea of a deity outside the Universe"[4] as a unique concept among the world's early religions. Unlike every contemporary religious writing, the Bible did not allow images to be made of God, as if He were merely a physical God who belonged to this world. Not only did Solomon admit that his temple was incapable of holding Him, but that heaven and the heaven of heavens could not contain Him (1 Kings 8:27).

The First Cause Must Be All-Knowing (Omniscient).

It is reasonable to assume that the Creator of all that is knows all about His own creation. In recent history humans have only begun to appreciate the complexities of the atom, of DNA, of the symmetry and harmony of nature's laws. As Einstein said, "the harmony of natural law . . . reveals an intelligence of such superiority that, compared with it, all the systematic thinking and acting of human beings is an utterly insignificant reflection."[5]

Scientists have come to expect a unified framework for nature's laws because all our experience in discovering them shows that they work together with tremendous precision (which physicists usually call "fine-tuning") to make life possible. There is a supreme rationale behind them. The laws of the universe yield evidence of perfect forethought, not arbitrary patchwork. When I questioned today's foremost theoretical physicist, Stephen Hawking, about why today's physicists expect to find some sort of grand unification for all of physics, he told me: "If the universe is governed by rational laws, which I believe it is, these laws shouldn't be an arbitrary patchwork, but should fit together into some unified framework."[6]

Surely these rational laws are fully comprehended by the One who set them up, even if science is not yet able to uncover all their mysteries (but merely able to find strong hints of their rational and unified nature).

The First Cause Must Have Personhood.

And last, though this God is spirit, we may logically infer that He is more than some amorphous, purposeless blob, because of the obvious purpose, the will, shown in His tremendous design of the universe, as we'll see in Chapter 9. The more physicists learn about the universe, the more they come to appreciate how impossible it would be for all the right conditions for life to come together in our universe by chance. A universe with stars and life requires a precise balance between the strength of nature's four fundamental forces, a very precise ratio of proton to electron mass, etc. Stephen Hawking describes how the values of the many fundamental numbers in nature's laws "seem to have been very finely *adjusted* to make possible the development of life,"[7] and how God appears to have "very carefully *chosen*" the initial configuration of the universe.[8] "Carefully choosing" and "finely adjusting" are clearly acts of the will.

Other prominent scientists, from Fred Hoyle[9] to Heinz Pagels[10] (though agnostics themselves) have also written of the clearly evidenced *purpose* or *intention* to bring about intelligent life in the universe. *Purpose* is perhaps the most important attribute of personhood, and the fact that the Grand Designer has this attribute suggests that He possesses personhood in some form.

Certainly this makes more sense than to think that personhood—the highest state of mind we know—is something that we possess but that God lacks. Again, because reason and observation tell us that every cause is always at least as great as its effect, we should expect that the First Cause of persons should Itself possess at least as much of all the attributes of thought, will, emotion, etc. that constitute our personhood (see Psalm 94:9, Genesis 6:6, Isaiah 65:19). Once again, this concept agrees well with the Bible, which emphasizes that we persons were created in His own image (Genesis 1:27).

Such a view provides clear answers to questions that perplex those with unbiblical viewpoints. Astronomer Fred Hoyle cites a modern theologian who poses a question which Hoyle considers unanswerable: "What we cannot understand is that God who has no need of the world should have reason to create (it)."[11] This is indeed a problem for Hoyle, who tries to conceive of God as being somehow less than a person, yet an "intelligence" that has left its obvious mark of "purpose in the Universe."[12]*

This "unanswerable" question has a ready answer for those who accept the Bible's description of a God who made humans in His own image, who has all the attributes of personhood, including a will. Such a God has no needs—and yet has *desires*. He had no need to create the universe or us in order to sustain His existence, but it *pleased* Him to do so. Thus our very existence is evidence for a personal God, a God who has desires, as opposed to an impersonal force.

And so our original proposition is not unreasonable or as difficult to support as some might first think. Using logic alone, it appears that the God

---

*Recognizing the need for a Supreme Being, but refusing to consider the idea of a biblical God, many scientists find pantheism a comfortable alternative. As you will see, many express this core belief by routinely capitalizing the words *universe* and *cosmos*.

Steven Weinberg on God

Steven Weinberg received the 1979 Nobel Prize for physics for his work in unifying two of nature's fundamental forces: the electromagnetic and weak nuclear forces. His classic book, *The First Three Minutes*, presented a detailed explanation of the earliest events in our universe, and it had a great influence on today's generation of big bang cosmologists. Weinberg hovers somewhere between agnosticism and atheism. Yet in his most recent book, *Dreams of a Final Theory: The Search for the Fundamental Laws of Nature*, he shows how, if there *is* a God, it is meaningless to think about Him merely as "our better nature" or "the universe" or "energy." He continues:

> It seems to me that if the word "God" is to be of any use, it should be taken to mean an interested God, a creator and lawgiver who has established not only the laws of nature and the universe but also standards of good and evil, some personality that is concerned with our actions, something in short that it is appropriate for us to worship. This is the God that has mattered to men and women throughout history.[13]

of the Bible may provide the best explanation we have for the origin of the universe. Reason tells us not only that the universe must have had a First Cause, but that the cause must be *independent* of its creation, like the God of Genesis 1:1; that it must have *infinite capabilities*, like the Almighty in the Bible; that it must *transcend time*, like the "I AM" of the Bible; that it must *transcend the physical*, like the Spirit who can only be worshipped in spirit and in truth; and last, that it must *possess a will to perform tremendous undertakings on behalf of us small creatures*, which is evident not only from natural revelation—in the way the universe was very precisely created for the benefit of human life (see Chapter 9)—but most of all, for those who take it seriously, from the Bible's supernatural revelation. Here we learn of how God Himself took on human flesh to die for us, demonstrating in the most extreme way possible His will, His purpose, in fact, His *love* for us.

Shocking Summary Statements and Stimulating Conversation Starters

◆ The idea that the entire universe came from nothing is irrational, not just to the Western mind, but to the *human* mind.

◆ Logic demands a cause for every effect. The universe is an effect that demands a very great Cause.

◆ Logic tells us not only that there must be a Creator, but it also tells us a good deal about what kind of a Creator He must be.

◆ The First Cause must be independent of its effect, that is, separate from the universe. From ancient times, the Bible has presented God as non-physical, a Spirit who cannot be contained, even by the heavens. Unlike other ancient religious writings, the Bible did not allow images of God to be made, as if He had a physical body.

◆ Nothing that is confined to time could have created the universe. From ancient times, the Bible has specifically defined God as the I AM, operating outside of time and existing before the creation of the universe.

◆ Though not a believer, Spinoza reasoned that the First Cause must be limitless in capability, like the "Almighty" in the Bible.

◆ The Bible also presents us with a God who is all-knowing. Einstein recognized that "the harmony of natural law . . . reveals an intelligence of such superiority that, compared with it, all the systematic thinking and acting of human beings is an utterly insignificant reflection."

◆ The finely tuned design of the universe requires a Creator with tremendous purpose, and thus, personhood. Believing that God is a person certainly makes more sense than to believe this highest state of mind we know is something that we possess but that God lacks.

The kind of care and perfection scientists observe in the universe's fine tuning points to exactly the kind of God described in
the Bible: a God who would go to the greatest lengths we can
imagine because of His love for us, but whose perfection would
prohibit Him from simply overlooking all the evil in our world.

Notes for Chapter 4

1. Martin Heidegger, quoted by Heinz R. Pagels, *Perfect Symmetry* (New York: Simon and Schuster, 1985), p. 137.

2. Ludwig Wittgenstein, quoted by Pagels, p. 137.

3. John Romer, *Testament* (television documentary series), aired on PBS in January, 1991.

4. Fred Hoyle, *The Intelligent Universe* (New York: Holt, Rinehard and Winston, 1983), p. 249.

5. Albert Einstein, *Ideas and Opinions—The World As I See It* (New York: Bonanza Books, 1974), p. 40.

6. Stephen Hawking, in letter to the author, July 12, 1994.

7. Stephen W. Hawking, *A Brief History of Time—From the Big Bang to Black Holes* (New York and others: Bantam Books, 1988), p. 125. Emphasis added.

8. Ibid., pp. 122, 127. Emphasis added.

9. Hoyle, pp. 217, 218.

10. Pagels, pp. 353-360.

11. Hoyle, p. 249.

12. Ibid., p. 217, 218.

13. Steven Weinberg, *Dreams of a Final Theory—The Search for the Fundamental Laws of Nature* (New York: Pantheon Books, 1992), p. 244.

Virgo cluster of galaxies. —*National Optical Astronomy Observatories*

CHAPTER 5

The Alternatives:
The Non-God Explanations

The fool says in his heart, "There is no God."
Psalm 14:1a

Carl:	Do you think people are ready for the cosmetology now?
Fred:	That's *cosmology.* The origin of the universe, not the origin of pimples.
Carl:	Pimples. Now there's a subject that will capture the attention of the younger market. But except for physics majors, how many people are really interested in going into all the scientific details about the origin and structure of the universe? Let's face it—cosmology isn't for everyone.
Fred:	No, not quite everyone. It's not for Greek gods, who have no concern about their mortality.
Carl:	That's right, Greek gods don't—hmm?
Fred:	It's not for angels or demons, for the same reason. It's not for swamis who have become one with the universe and already know all its secrets. Or for certain preachers who have already memorized all the "proper" answers.
Carl:	So who's cosmology for?
Fred:	Cosmology is for regular folks. Cosmology is the natural, preordained quest of *Homo sapiens*—of every rational person.
Carl:	How do you figure?
Fred:	When people—people everywhere, in every age—look up into a starry night sky, they can't help but to start *wondering*. There's something about looking up at the stars that has an effect on people—unlike the effect it has on apes or dolphins or dogs or whatever animal you think is intelligent. Because only *people* can *con-*

> *template* the stars. And we can't contemplate them for very long
> without wondering: Where, oh where, did all this come from?
>
> *Carl:*   Obviously a very great power. Probably even bigger than the Tri-
>   lateral commission.
>
> *Fred:*   The philosophers put it this way: "Why is there something instead
>   of nothing?" Coming to grips with how everything got here and
>   how *we* got here and *why* we're here and whether there's a God
>   and if so *Who:* this is the human's job. Nothing could give us more
>   lasting satisfaction than answering these questions.
>
> *Carl:*   But we can't, can we? I mean, not without the Bible.
>
> *Fred:*   Some people don't believe the Bible.
>
> *Carl:*   So you start with science.
>
> *Fred:*   You can. You can even start with all the beliefs and theories that
>   people have proposed to say that maybe there *is* no God. Then
>   you're in a position to see exactly what the alternatives to God *are*.

A Skeptic's Questions:

Even if common sense tells you that the universe had to have a great First
Cause, sometimes common sense is wrong. Science might have another
explanation. It's still possible that the universe has simply always been here,
like the Eastern religions say. In fact, there are lots of scientific theories that
say the same. There's the steady state theory, the plasma theory, cyclic theories—any one of them could completely destroy your argument for a
Creator who's separate from nature and spiritual and eternal and all that.

A Bible Believer's Response:

Let's take a look at each of those theories, and you tell me which one has a
better chance of being true than the Bible's teaching that in the beginning
God created the heavens and the earth.

The Universe Without a Start

Eastern Explanations

Some Eastern religions say that the universe has simply always *been,*
that the universe is its own cause. The cause and the effect are the same.
God and the universe are one. "There is no duality," as the swamis say.

Again, this contradicts the obvious principle from logic that every cause demands an effect. Contrary to much popular belief (which stems perhaps from a good-intentioned but misplaced effort to show acceptance of other cultures), this logical principle is not just a Western notion—it is common to human understanding. People of any region of the world must be indoctrinated with a belief system contrary to their own logic in order to accept the idea that a cause and an effect can be the same. The natural belief in the separateness of God from His creation must be forcefully overcome. This is evident from the fact that "duality" must be continually taught to be an "illusion."

Some readers might find it interesting to hear Stephen Hawking's response when a science writer tried to connect his findings on black holes with certain Hindu myths: "It's fashionable rubbish," Hawking said. "People go overboard on Eastern mysticism simply because it's something different that they haven't met before. But as a natural description of reality, it fails abysmally to produce results."[1]

The Upshot of Subjective Thinking

Of course, Eastern thought has had its influence on Western society in the 20th century. Nominal Christians and non-observant Jews have been most susceptible to converting to a Bahai-like, there-are-many-ways-to-God mentality. Since we're all God's children, let's not worry about sin and grace and heaven and hell—let's just all get together and feel good about ourselves. Let's avoid any unpleasantness that might divide us and concentrate on reaching our full, feel-good potential. Truth ends up taking a back seat to "self-realization."

Physicist Steven Weinberg has little use for religious fundamentalists. Yet he writes:

Religious liberals are in one sense even farther in spirit from scientists than are fundamentalists and other religious conservatives. At least the conservatives like the scientists tell you that they believe in what they believe because it is true, rather than because it makes them good or happy. Many religious liberals today seem to think that different people can believe in different mutually exclusive things without any of them being wrong, as long as their beliefs "work for them."[2]

In their pursuit of truth, scientists and religious conservatives share the same spirit. Both believe in an objective reality that is the same for all, not alternative realities that can be chosen according to what works best for the individual. In this spirit, a Christian should believe the Bible because it is true; he should not believe that the Bible becomes true because he believes it.

Yet, to their discredit, some conservative Christians have more in common with religious liberals and Eastern thinkers than they do with scientists. By starting their belief system with an irrational commitment to particular traditions or biblical interpretations, they often come into unnecessary conflict with science. As long as their particular interpretations help them feel happy or good about themselves, they will allow these traditions to preclude their consideration of plain scientific evidence. Some even end up associating most of the scientific community with the work of the devil.

For their part, many agnostic scientists prefer to see science and religion thus placed in separate compartments far out of reach from one another, rather than acknowledging that each has a right to explore different aspects of the same reality. The more foolish the religious folks appear, the more secure the agnostics can feel about their naturalistic world view. We'll have more to say about the relationship between science and faith later.

Steady State Cosmology

The scientific equivalent of Eastern religious thinking was once the "static universe," long favored by many scientists because it dodged questions about ultimate origins. When astronomical evidence for an expanding universe became undeniable, an infinitely old universe became impossible unless some accounting could be given for the fact that we still had billions of galaxies in sight. In an infinitely old, expanding universe, they should have separated to remote distances long ago.

From 1918 on, new theories attempted to suggest various means whereby matter might be continuously created in order to account for the existence of nearby galaxies in an expanding universe. These were based on various notions of how radiation or other forms of energy might be converted into matter and eventually coalesce into new stars.

The most widely-accepted version of this idea, the steady state theory, actually began with a ghost story. It seems that the steady state's three originators—physicists Hermann Bondi, Thomas Gold, and Fred Hoyle—went out together to see a horror film one night in 1946, and the ghost story inspired them. The movie's plot was circular, so that it cleverly ended at the same place it began. Over brandy later that night in Bondi's Cambridge apartment, they wondered if the universe might be something like the movie, the end continually being replaced by the beginning.[3]

Hoyle's first paper on the new theory was apparently rather skimpy; its publication was rejected, as he described it, "on account of the acute shortage of paper."[4] Once fleshed out, the theory proposed that a "creation field" be added to Einstein's general relativity equations to show that matter was somehow being created at a rate that precisely counterbalanced the universe's expansion, so that the average density of the universe remained constant. All forms of the theory stated that matter was continually being created in the expanding space between galaxies, but no mechanism was ever found to account for this creation.

The steady state theory has now fallen far out of favor with almost all scientists who are concerned with cosmology, especially as a result of astronomical observations cited in the next chapter. Einstein initially preferred the "static universe" explanation and felt so strongly about it that he added a fudge factor, officially called his "cosmological constant," into his relativity equations in order to preserve the idea of a universe without a beginning. Contrary mounting evidence, especially from astronomy, led Einstein to admit this addition as a great mistake[5] and to hold, as nearly all cosmologists now do, that the universe must have had a beginning.

Sir Fred Hoyle, the steady state theory's greatest promoter, brought about the demise of his own theory in 1964 when he collaborated with R.J. Tayler at Cambridge in an attempt to prove that the stars alone could have produced the amount of observed helium in the universe today. Astronomers have learned from spectroscopic studies of starlight that about three-quarters of our universe is composed of hydrogen (the lightest element), about one-quarter is helium, and a tiny percentage is comprised of all the rest of the heavier elements. Though stars are capable of leaving behind the required *heavier* elements, Hoyle demonstrated that even if a star had been burning for the entire age of the universe, there still would not have been enough time for it to convert more than about 3 percent of its hydrogen

Sir Fred Hoyle, independent-thinking astrophysicist, in 1974. —*Courtesy of the Mary Lea Shane Archives of the Lick Observatory*

into helium. The big bang theory, however, did perfectly predict that enough hydrogen should have been converted to helium in the initial high density state so that about 25 percent of the universe should be composed of helium today. In Hoyle's own words:

> We found ourselves convinced that all the matter in the Universe must have emerged from a state of high density and high pressure . . . . Our results, together with further developments by William Fowler, Robert Wagoner and myself, became what even to this day is pretty well the strongest evidence for the big bang, particularly as the arguments were produced by members of what was seen as the steady state camp.[6]

These results notwithstanding, Hoyle is one of the very few scientists who continues to stubbornly search for ways to propose an eternal universe, since a universe with a beginning requires an outside agency that cannot be explored by science. In particular he has tried to make his theories more palatable (though unsuccessfully) to his colleagues by hypothesizing that perhaps "little big bangs" have produced the galaxies—*within* the uni-verse—rather than demanding that one big bang be responsible for the entire universe.[7]

An insurmountable problem for this idea (and for all models of a steady state theory) is the simple fact that all galaxies in the vicinity of our galaxy are middle-aged. None are found in any stages of early formation, as steady state models would predict.

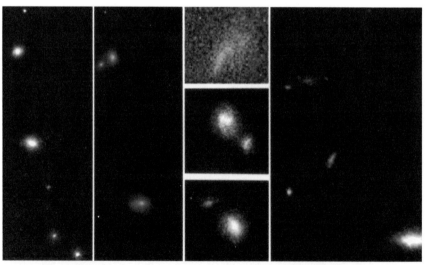

According to NASA, the Hubble Space Telescope's Deep-Sky Survey reveals irreg-ular shapes for galaxies over three billion light-years distant. They don't fit the familiar pattern of nearby spiral and elliptical galaxies, and some seem to be inter-acting. —*Courtesy of NASA*

Nearer galaxies take the more familiar spiral and elliptical shapes, as in Stephan's Quintet, five galaxies (four shown) in the constellation Pegasus. As seen here, some relatively nearby galaxies interact too, but this is much rarer among closer galaxies. Four of the five members are gravitationally related and have identical redshifts, showing them to be about a quarter of a billion light-years away. —*National Optical Astronomy Observatories*

However, given the fact that distant galaxies are all receding from us, if the universe was created at a point in time in the finite past, we might expect to see younger-looking galaxies at greater distances. Since the light from the farthest galaxies takes billions of years to reach us, the farther we look into space, the further we look back into time. And indeed, as Princeton cosmologist Jim Peebles reports, "the radio galaxies observed [at greater distances] tend to look immature, with irregular and extended optical shapes. This is consistent with their youth . . . ."[8] Also, astronomers have found that distant (earlier) galaxies are distributed more closely together than are nearby galaxies. And at the greatest distances (that is, in the early universe), a different type of celestial object becomes prominent—quasars—which many believe are the precursors of galaxies. We will examine these observations in greater detail in coming chapters.

Fred: So, Carl, what's your opinion of the book so far?

Carl: It's worthless.

Fred: I know, but I'd like to hear it anyway.

Carl: (Sigh) You're hilarious. I just think the book has nothing to offer the average person. It's getting too far over people's heads. You need to get more down to earth.

Fred: Well, we're studying the heavens. We'll get down to earth when we study geology in Volume 2.

Carl: But all this talk about spectroscopic studies and converting hydrogen into helium—how am I supposed to follow that?

Fred: When stars burn, they convert hydrogen into helium. That's not so hard to understand, is it?

Carl: Maybe, but how's a skeptic supposed to know that really happens?

Fred: We can measure the amount of helium in a star by using a prism or filter to break down its light into its constituent colors. Each element has its own spectrum.

Carl: This is beginning to sound like a high school science course.

Fred: If you can just hang on a little longer, I'll start explaining more of these terms with concrete examples. Right now, remember, we're just talking about the alternatives to God and to a universe with a beginning. If I stopped to explain everything in detail now, we'd really get bogged down.

Carl: I could be watching television right now.

Fred: Yes, but then you'd never find out the latest dirt on plasma cosmology.

Plasma Cosmology

This not-so-popular challenge to big bang cosmology proposes that most of the universe is composed of plasma (electrically conducting gases). In one of the versions of the theory proposed by the father of plasma cosmology, Hannes Alfvén, the plasma indirectly produces a repelling effect between galaxy superclusters, causing the observed expansion of the universe. However, rather than starting this expansion from a singular point, as in big bang cosmology, Alfvén proposes a sort of partial big bang. The universe expands and collapses, but instead of contracting to a point, it contracts to a size about one-hundredth of its present size. Then some as-yet-unknown cosmological principle kicks in to blow the universe apart once again, thus maintaining eternal equilibrium.[9]

In contradiction to the second law of thermodynamics, the universe never wears out, but merely recyles its old forms of energy to continually find new sources. When all the hydrogen is one day burned into iron and fusion is no longer possible, the universe will inevitably find a new source to energize it. Why, we might ask, is such a process inevitable? Because the theorists have begun with the presupposition that the universe is eternal, and some sort of continual recycling seems necessary to account for the fact that the universe didn't run down a long time ago.[10]

The theory does not readily account for the observed amounts of helium and the light isotopes observed in the universe (which cannot have been synthesized in sufficient quantities in stars alone, but which *are* explained by the big bang). Its attempts to find alternative explanations for the other phenomena that support the big bang theory (the microwave background radiation and the Hubble expansion, discussed in chapter 7) remain highly speculative. Heavy antimatter particles, which should be relatively common according to Alfvén's partial-bang hypothesis, have been sought but never found.

The healthiest thing about plasma cosmology, from a healthy skeptic's point of view, is that it calls into question all the recent findings of big bang cosmology and so may encourage science to make a healthy re-examination of its assumptions.

More important to our subject: plasma cosmology, like steady state cosmology, gives no explanation for ultimate origins. Plasma popularizer Eric Lerner proposes a "starting place" for the universe when it was "filled with a more or less uniform hydrogen plasma, free electrons and protons."[11] Concerning what brought this plasma into being, he writes that "we have no real knowledge of what such processes were."[12]

The Universe That Starts—Without a Starter

No position is more difficult to defend than one that tries to combine the big bang theory with atheism or pantheism. This is why, though the big bang theory today has a strong consensus among scientists, it was so abhorred by the scientific community at first. It didn't fit their non-biblical preconceptions.

British astrophysicist Arthur Eddington wrote, "Philosophically, the notion of a beginning of the present order of Nature is repugnant to me. . . . I should like to find a genuine loophole."[13] Einstein tried to avoid such a beginning by creating and holding onto his cosmological "fudge factor" in his equations until 1931, when Hubble's astronomical observations caused him to grudgingly accept "the necessity for a beginning."[14] We have already seen how Hoyle dislikes the idea because, as he puts it, "The big bang theory requires a recent origin of the Universe that openly invites the concept of creation."[15] Barry Parker sums up the feelings of most cosmologists: "If we accept the big bang theory, and most cosmologists now do, then a 'creation' of some sort is forced upon us."[16]

Obviously, the problem for those with a nontheistic perspective is to come up with a way by which the universe could start from truly nothing without anything or anyone to start it. Stephen Hawking, widely regarded as the most brilliant theoretical physicist since Einstein, has diligently applied himself to the task; and although he has come up with a mathematical proposal to suggest how time and space might be finite but without boundary, he calls this mathematical model "imaginary time"[17] and emphasizes that this is "just a *proposal:* it cannot be deduced from some other principle."[18] "In real time," he says, "the universe has a beginning . . . ."[19]

Stephen Hawking and George Smoot at a conference at Lawrence Berkeley Laboratory. Because he is afflicted with a motor neuron disease, known as Lou Gehrig's disease or amyotrophic lateral sclerosis, Hawking has lost most voluntary motor functions. After pneumonia led to a tracheotomy in 1985, he lost all ability to speak. Hawking now communicates through a computer-activated voice synthesizer, selecting words one by one with two fingers. He quips that he has now become IBM-compatible and asks that his listeners please pardon his American accent. —*Courtesy of Lawrence Berkeley Laboratory, University of California*

Hawking's Proposal:
What if There Was No Beginning?

In his introduction to Stephen Hawking's *A Brief History of Time*, Carl Sagan says that Hawking has reached the tentative "conclusion" that our universe has "nothing for a Creator to do"; he says the book is "perhaps about the absence of God."[20] However, the book makes it quite clear that Hawking carefully distinguishes between "the real time in which we live"[21] (which apparently contains black holes and a beginning) and his mathematical proposal, which by definition speculates on the possibility of no boundaries in space or time—and hence no beginning. Hawking himself claims no conclusion about the absence of a Creator. In fact, in his correspondence with me, he has told me that even if his proposal turned out to describe the real universe, no conclusions could be drawn about God's absence, but rather about His nature:

> I do not believe the no-boundary proposal proves the nonexistence of God, but it may affect our ideas of the nature of God. We do not need someone to light the blue touch paper of the universe.[22]

Even a universe without a beginning does not necessarily result in a universe with nothing for a Creator to do, for there is apparently more to running our universe than merely igniting it in a big bang. Rather than limiting God's involvement to winding up the cosmos and just letting it go (as the old deists believed), perhaps we should give more consideration to the need for God's involvement in holding all things together *now* (Colossians 1:17). From a biblical perspective, we must agree that it is even more in God's nature to be the ruler of the universe than to be the mere initiator of it.

Of course, I must finally stress, as Hawking does, that his proposal is merely a mathematical one: "it may be put forward for aesthetic or metaphysical reasons, but the real test is whether it makes predictions that agree with observation."[23] Notwithstanding all the hoopla this proposal has generated, scientists have no theory to show how such a universe without boundaries might exist in reality. How can we combine our knowledge of an expanding universe with a no-boundary condition for space and time? No one has yet proposed a physical explanation to go with the mathematical model.

Alan Guth, father of the inflationary model of the big bang theory, told me that Hawking's proposal "suffers from the problem that it doesn't yet have a completely well-defined theory in which to embed it. That is, it really is a notion of quantum gravity, and so far we do not have a complete quantum theory of gravity in which to embed this idea."[24]

As we will see in greater detail in the next chapter, Einstein tried to find an explanation for his general relativity equations that would not require a beginning and a Beginner for the universe, and he came away a believer in both. He later wrote of his desire "to know how God created the universe."[25] We must conclude that when the world's greatest minds are put to the task of finding a way to propose a universe without a Creator, they find the task impossible, at least as far as can be shown by any theory that would line up with what we know of the real universe.

Combinations of Starts and Non-Starts

The Steady State Becomes Unsteady

A serious problem arose soon after the big bang theory was put forward: the earth appeared to be older than the proposed age of the universe. Calculations made the big bang more recent than the formation of the earth's crust, a time known to geologists from the radioactive decay of our oldest rocks. Obviously the earth could not be older than the rest of the universe.

To explain how the universe might appear younger than it is, Belgian physicist Georges Lemaître proposed that the universe could have exploded out of the big bang, slowed itself by gravitational attraction to a near stop for an indefinite period of time—held there by a hypothetical anti-gravity force—and then continued its expansion when this force eventually became dominant. However, no explanation could be found for any such "anti-gravity" force or the tremendous precision it would have required to bring an expanding universe to a standstill.

Because of Arthur Eddington's disdain for beginnings, he took advantage of the opportunity to return to an infinitely old universe and developed a model that began in a static state, became unstable, and then began to expand according to our present observations of receding galaxies. But this presented him with more problems, since he himself had shown that Einstein's general relativity equations result in an unstable universe: one that must either expand or contract. The static beginning could not be maintained.

When better astronomical data expanded the distances to the farthest galaxies and hence the age of the universe, this cosmic hesitation theory became unnecessary and was discarded.

Quasi-Steady State Cosmology

During 1993 and 1994, Hoyle was joined by Geoffrey Burbidge and

Jayant Narlikar in an attempt to resurrect Hoyle's mini-bang theory, calling it "quasi-steady state cosmology" (QSSC). According to QSSC, the proposed "creation field" (from which new matter springs) only exists in certain regions of high mass density. The strength of the creation fields alternately increases and decreases during the history of the universe, resulting in alternating cycles of slow and rapid expansion. The universe, they say, is now about a trillion years old, and we happen to be living in the middle of a short-term slowdown in its expansion, giving it the appearance of perfect balance between gravitational collapse and an expansion that would be too rapid to allow galaxies to form.

Unhappily for the trio, they have not yet come up with a convincing explanation for the smooth microwave background radiation. A single, universal explosion perfectly explains this extremely smooth radiation now observed coming from every point in the sky; a series of widely-separated mini-bangs does not. Moreover, scientists can find no basis from physics for the proposed creation fields.[26] The notion is a purely speculative, philosophical one, though the team claims that their theory can explain certain other cosmic realities (the temperature of the microwave background radiation, the abundance of light elements in the universe, etc.). Physicist David Schramm of the University of Chicago says: "They have so many parameters that they tweak and play around with that they can fit anything, but it's not a real fit."[27]

Even if we accept QSSC and its trillion-year-old universe, we are no nearer to an explanation for a beginning. QSSC puts off the question of cosmic origins further into the past, but it still fails to tell us anything about how something could come from nothing.

Carl: Creation fields? Microwave background radiation? Am I supposed to be following all this scientific jargon?

Fred: Think of a creation field as a football field in which new football players suddenly pop on the field out of nowhere.

Carl: That's impossible.

Fred: That's the point. And think of the microwave background radiation as the heat in your microwave oven. You turn it off and open the door and the left-over heat rushes out. But what if we lived in the oven and the door was stuck shut?

Carl: We'd get cooked.

Fred: But what if we lived in a special oven that kept getting bigger?

Carl: I guess the heat would spread out and it wouldn't be so hot when the oven got big enough.

Fred: That's right. Just like the microwave radiation has now settled down to just 2.7 degrees Kelvin—above absolute zero—everywhere we look. And we'd expect the heat to spread out evenly, because the space inside the oven all started at the same temperature.

Carl: Well, I still don't see anything so funny about it.

Fred: What?

Carl: The comic background radiation.

Fred: That's *cosmic* background radiation. It just means it's coming from everywhere in the universe. You'll see where all this is heading if you can just hang on till we get into the next couple of chapters.

Cyclic Cosmology

According to Einstein's general relativity equations, whether the universe continues to expand or eventually reverses itself and contracts depends upon the total mass within the volume of our universe. If the density of matter in our universe exceeds a critical amount, all the matter in the universe will eventually converge back together into a "big squeeze." This, of

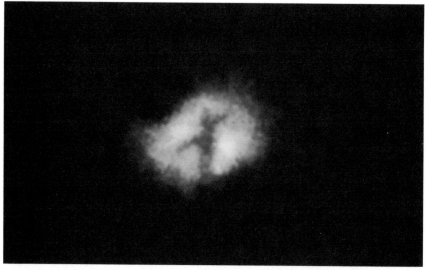

"X marks the spot" of a probable black hole in the nucleus of the nearby spiral Whirlpool Galaxy. With a mass equivalent of about one million suns, the black hole may be what is attracting two swirling dust disks, seen edge-on. This photo was taken from the Hubble Space Telescope. —*Courtesy of NASA*

course, has led some cosmologists to wonder if perhaps another big bang might subsequently take place and a new universe might emerge from it, and perhaps the process will eternally repeat itself. This "oscillating universe" or "re-bang" hypothesis has a precedent in the cyclic worlds proposed by the ancient Stoics and even earlier by the Hindus, and it has something in common with the philosophical notion of eternal reoccurrence.

However, cyclic cosmology still offers no explanation for how the process began. Though some might use it to postpone the question of *beginnings*, they still can't avoid the need for an ultimate *cause*.

The actual physics of a cyclic universe poses serious problems for anyone who wishes to take it seriously. According to Einstein's general theory of relativity, once matter crunches into a singularity (as in a black hole), nothing, not even light, can escape. Black holes possess an "event horizon" which acts as a one-way membrane; matter can enter, but never leave.

What's a Black Hole?

The idea of a black hole can be understood in terms of the escape velocity that is associated with all gravitating bodies. In the case of earth, a projectile must reach a speed of 25,000 miles per hour before it can escape earth's gravity. If a body is massive enough, as in the case of some neutron stars, even the tightly packed neutrons would give way and the star must collapse under its own gravity into a "singularity," a point of zero volume. According to general relativity, this means that the density of its matter and the curvature of space-time surrounding it become infinite. Not even light can escape this infinite curvature—so it is a "black hole." The escape velocity rises beyond the velocity of light. Since nothing can travel faster than light, nothing can escape (although Stephen Hawking's studies of black holes permit them to leak gamma rays and eventually dissipate).

Do black holes really exist? In the early 1970s, Hawking officially bet Caltech's Kip Thorne that Cygnus X-1 would eventually be shown *not* to be a black hole. This blue supergiant is observed to be whirling around an invisible object every five days. Whatever it is, Cygnus X-1 must be incredibly massive. As Hawking admitted, "If it isn't a black hole it really has to be something even more exotic."[28] By 1990, alternatives to the black hole had been ruled out—at least to Hawking's satisfaction. He conceded that he had lost his long-standing wager and paid up.

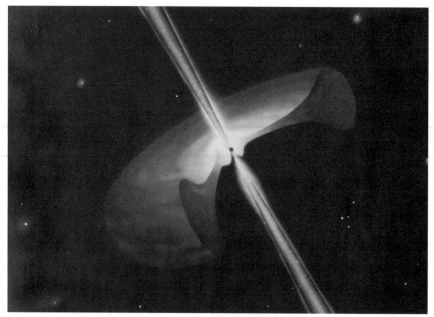

Gamma ray bursts may be evidence of black holes, formed from collapsed stars (also called "failed supernovae"). If one of these beams of energy falls in our line of sight, this would explain the tremendous power of observed gamma rays. —*Courtesy of NASA*

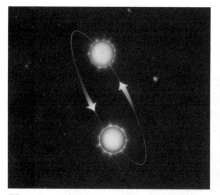 

Here is an alternate explanation for the tremendous energy of observed gamma ray bursts: Two orbiting neutron stars merge, producing beams of energy, one of which falls in our line of sight. —*Courtesy of NASA*

If matter cannot re-bang out of a black hole singularity, then neither could it re-bang out of a universal singularity. And even if quantum mechanics were to some day give us an explanation for how matter might become "untrapped" from a singularity to produce our big bang, physicists find that the result would not be a regular cycle of expansion and contraction, but of ever greater and more chaotic expansions, which do not explain the orderly universe in which we find ourselves.[29] Physicist Richard Morris also writes that such theoretical cycles must have been shorter in the past. Thus:

> It is possible to calculate that in this case there can have been at most a hundred bounces. In other words, the universe must have had a beginning at a finite time in the past. *We are thus led back to the 'problem' of creation out of nothing that the oscillating universe was designed to avoid in the first place.*[30]

Alternatives to Creation

Interview with George Smoot

George Smoot served as leader of the COBE (pronounced "coby") satellite team that first detected the long-sought ripples in the cosmic background radiation. This finding provided observational evidence to help decide between theories (to be discussed later). He made the following remarks about the modern consensus on alternative cosmologies.[31]

Heeren: How highly regarded now are alternative cosmologies by the majority of astrophysicists—steady state or quasi-steady state theories, plasma cosmology, and so on? Are they dead, or are there still some people out there holding to them?

The Cosmic Background Explorer (COBE) satellite. —*Courtesy of NASA*

Smoot:	There are still some people out there holding to them. I just got the stuff for a meeting that I'm going to give a talk at. As it turns out, both Hoyle and Burbidge are going to give talks at it. Hoyle is one of the original steady state persons. And Burbidge is not far behind. And they have a new mini-bang— a quasi-steady state model—which is a series of little-sized, middle-sized bangs. And there's still a disciple of Hannes Alfvén, named Eric Lerner, who pushes the plasma cosmology, but if you ask, what is the situation going to be like in a few years, when Hoyle and Burbidge and Narlikar and those guys die or retire, there won't be many people supporting the theory. And in the case of Eric Lerner, he's pretty much out there by himself.
Heeren:	Right. Is Hannes Alfvén still alive?
Smoot:	I think he is. But he doesn't go out and propose alternate cosmologies right now.
Heeren:	So when those guys go, there aren't going to be many people around to support it.
Smoot:	Right. Well, Lerner is much younger. So he should be around for awhile. But I certainly don't think any alternative cosmologies have any following.

There Simply Are No Natural Explanations for Ultimate Origins

We have seen that science provides just two fundamental alternatives for the formative history of the universe—either it has always been here or it began at some point in the past—but neither gives an explanation for ultimate origins. Both explanations fail to answer the greatest question of all: Where did matter and energy come from?

The Limits of Science

Modern science has no laws or observations to show how something could have come from nothing; it doesn't even have a *theory* to propose such an event. It is a chief tenet of science that while it studies *formation,* ultimate *origins* fall outside of its domain.[32] Also, for very practical reasons, ultimate origins are impossible for science to study since, as physicist Milton Rothman points out, "there is no way of obtaining evidence concerning a prior existence" (i.e., before the big bang).[33] All the laws of physics—even the laws of general relativity—break down before "Planck time" (explained below).

The Limits of Scientists

Those scientists who claim that *science* tells them something about ultimate origins are not being quite honest, be they atheists or creationists. Science is limited by what can be known from observation, which big bang cosmologists admit cannot be extended back before the time known as Planck time, when all the laws of nature break down near the point of singularity (the big bang).

Before this Planck moment (named after Max Planck, a pioneer of quantum physics), the particles we know today could not have existed. According to modern physics, this point occurred 10^{-43} seconds after the big bang,* when the universe was an unbelievably hot point smaller than an atom. Princeton's influential cosmologist, Jeremiah Ostriker, told me:

> If you extrapolate back and back in time, the universe will become so dense that you will need quantum mechanics. And we know we don't have a quantum theory of gravity. At what time does this become a serious problem? At the Planck time. So we cannot extrapolate back our current theories before the Planck time.[34]

Physicist Barry Parker writes that most scientists neglect anything that happened before that time. But whether we neglect it or not, he says something must have preceded that moment:

> Unfortunately, a very critical event had happened—creation itself. And without a theory to explain this event we can only guess what happened. . . . How do we contemplate such a situation? . . . The only reasonable answer to this question is: we do not. Indeed, we cannot even make calculations describing it.[35]

Having stated this cosmological caveat, we can say that it is still perfectly ethical for a scientist or anyone else to state his metaphysical opinions about what caused or didn't cause the universe to reach that point. But it is wrong for anyone to say that *science* has provided him with the information, or for a scientist/celebrity to use his platform as a scientist to imply that science tells us anything about it.

Yet famed Cornell astronomer Carl Sagan confidently claims, "The Cosmos is all that is or ever was or ever will be."[36] In the same article in which Milton Rothman admits that "there is no way of obtaining evidence concerning a prior existence," he confidently implies scientific knowledge of that prior existence, stating that the big bang "started with a quantum fluctuation that created matter and energy at one point in time. Nothing went before."[37]

---

* For the extremely large or extremely small numbers required by cosmology, we must use power of ten notation. "10^{-43} second" denotes a fraction of a second which, if written out, would require a decimal point followed by 42 zeros followed by a 1.

Rothman offers no further explanation about what might have caused such a quantum fluctuation or how he knows that nothing went before. Scientists who have committed themselves to a non-biblical perspective have often used their platform as scientists to proclaim that the universe "could have come about by natural means," as astronomer Victor J. Stenger states, and that "the universe exploded out of nothingness."[38]

Carl Sagan. —*Courtesy of Carl Sagan*

After Sagan opened his book (and his television series) with the proclamation that the cosmos is all that there ever was, in all the pages (and hours) following, he gave no scientific reasons for making such a statement. In all his discussions of the big bang, he offered no theory to justify his faith in a cosmos that can explain its own existence. Sagan apparently mistakes the limitations of science for the limitations of reality. Science can't discover what happened before the big bang; therefore, he seems to reason, *nothing* could have happened before the big bang.

Many scientists, of course, take a broader view of reality. Arno Penzias, co-winner of the Nobel prize for physics for his discovery of the microwave background radiation, told me:

> We ought to make sure that, since scientists can only speak in physical terms, that they don't take that as being the entire world. . . . I think scientists are very poor witnesses, because they're looking at such a very small part of the world. That's my view. And they tend then to think of the physical part of the world which they're able to experience as all of reality.[39]

The closest thing to a scientific explanation for ultimate origins was summed up by well-known science (and science fiction) writer Isaac Asimov, and it is a view seriously resorted to by others:

> Where did the substance of the universe come from? . . . If $0 = +1 + (-1)$, then something which is 0 might just as well become 1 and -1. Perhaps in an infinite sea of nothingness, globs of positive and negative energy in equal-sized pairs are constantly forming, and after passing through evo-

lutionary changes, combining once more and vanishing. We are in one of these globs in the period of time between nothing and nothing, and wondering about it.[40]

A number of science popularizers have picked up on this explanation, since it uses known implications from quantum mechanics. Quantum theory says that space, though it appears to be empty, is actually filled with virtual particle pairs which may "fluctuate" or appear for extremely short periods of time. However, Ed Tryon, the physicist who originated the above much-quoted proposal in a 1973 article for *Nature*, apparently recognizes a problem with how even this theory could explain creation from true nothingness,[41] since the quantum effects he describes require something more than nothing—they require *space*, something all physicists now carefully distinguish from "nothing."

Fred Hoyle says, "The physical properties of the vacuum would still be needed, and this would be something."[42] The space in our universe is called a "false vacuum" because it contains properties that make it much more than "nothing"—it contains quantum particles and is not truly empty. Thus a false vacuum also demands a cause.

Physicist James Truran (University of Chicago) described to me how energy can be converted into protons and anti-protons, thus creating matter, but not from nothing: "You can create particle-anti-particles," he said, "but you have to have energy available to do it."[43]

Moreover, general relativity shows that the space in our universe is not just nothing. Einstein wrote: "There is no such thing as an empty space, i.e., a space without field. Space-time does not claim existence on its own, but only as a structural quality of the field."[44] Cosmologist Paul Davies points out that, to the physicist, when he asks about how something came from nothing, "that means not only, how did matter arise out of nothing, but 'why did space and time exist in the first place, that matter may emerge from them?'"[45]

When physicists speak of the expansion of the universe from a singularity, they are not speaking of an expansion within a larger space, but rather of space itself expanding with the big bang. There is nothing "outside" the universe. And yet it is this kind of nothingness from which the universe must have sprung. In this "true nothingness," nothing, including the phenomena of quantum physics, could happen.

John Mather (NASA's principal investigator of the cosmic background radiation's spectral curve with the COBE satellite) told me:

> We have equations that describe the transformation of one thing into another, but we have no equations whatever for creating space and time. And the concept doesn't even make any sense, in English. So I don't

think we have words or concepts to even think about creating something from nothing. And I certainly don't know of any work that seriously would explain it when it can't even state the concept.[46]

In other words, any kind of natural creation from absolutely nothing (*ex nihilo*), remains as unthinkable in this scientific age as it always has been for the philosophers. Some cosmologists would like to imply that they have found a way to explain this thorniest of cosmological realities, but all must stop short of claiming any true explanation for ultimate origins. The closest they can come to an explanation is to say that, given what we know of quantum fluctuations, something might come from *practically* nothing. George Smoot, principal investigator of the cosmic background's ripples with the COBE satellite, says: ". . . it is possible to envision creation of the universe from almost nothing—not nothing, but practically nothing. Almost creation *ex nihilo*, but not quite."[47]

When pressed to answer how creation could begin with *nothing*, Tryon himself (the originator of the fluctuation/creation proposal) said, "It may well be that we will never have a confident answer."[48] The earlier hope that something could have come from nothing now appears to be going the way of the "spontaneous generation" theory of the mid-1800s, the idea that life (even flies) could spontaneously spring from nonliving materials in a closed jar.

As it enters the 21st century, science has yet to come up with a natural explanation for the universe's origin, and it would seem that the supernatural explanation given in Hebrews 11:3 is still the best one we have: "By faith we understand that the universe was formed at God's command, so that what is seen was not made out of what was visible."

Clearly this kind of faith is at least as reasonable as any theory science can offer.

Conclusion: Science Ends Where the Bible Begins

Those scientists who have been honest about the question of where matter and energy originated have admitted two things: first, that the problem is impossible to solve through science, and second, that this state of affairs is exceedingly frustrating to the scientist.

Internationally respected astronomer (and self-confessed agnostic) Robert Jastrow admits that scientists have been "traumatized" by coming up against a problem that must forever remain beyond them.[49] In his book, *God and the Astronomers*, Jastrow says, "The development is unexpected

Robert Jastrow. —*Courtesy of Robert Jastrow*

because science has had such extraordinary success in tracing the chain of cause and effect backward in time."[50]

The situation violates a deeply held "religious faith" of scientists in science itself, the belief that science should eventually be able to discover the forces and laws to explain *everything*. After all, Carl Sagan tells us that science is "applicable to everything. With this tool we vanquish the impossible."[51] But Jastrow writes:

> Consider the enormity of the problem. Science has proven that the Universe exploded into being at a certain moment. It asks, What cause produced this effect? Who or what put the matter and energy into the Universe? . . . . And science cannot answer these questions, because, according to the astronomers, in the first moments of its existence the Universe was compressed to an extraordinary degree, and consumed by the heat of a fire beyond human imagination.[52]

Jastrow says that the universe began "under circumstances that seem to make it impossible—not just now, but *ever*—to find out what force or forces brought the world into being at that moment."[53]

Anticipating all such questions about the unknowable moment of creation, Isaiah tells us that no one can fathom the understanding of the Creator (Isaiah 40:28). But as to who or what is the cause for this effect, the Bible raises the question, Shouldn't we have known that answer all along? "Lift your eyes and look to the heavens: Who created all these?... Do you not know? Have you not heard? The LORD is the everlasting God, the Creator . . . ." (Isaiah 40:26a, 28a).

Compared to the alternative of supposing that matter and energy somehow always existed, British physicist Edmund Whittaker says, "It is simpler to postulate creation *ex nihilo*—Divine will constituting Nature from nothingness."[54]

Physicist Barry Parker agrees: "We do, of course, have an alternative. We could say that there was no creation, and that the universe has always been here. But this is even more difficult to accept than creation."[55]

After considering the discovery that our universe had a beginning and that science is incapable of ever discovering what went before, astronomer Jastrow concludes his book:

> For the scientist who has lived by his faith in the power of reason, the story ends like a bad dream. He has scaled the mountains of ignorance; he is about to conquer the highest peak; as he pulls himself over the final rock, he is greeted by a band of theologians who have been sitting there for centuries.[56]

Shocking Summary Statements and Stimulating Conversation Starters

⚡ Cosmology is the natural, preordained quest of every rational person. When we look up into a starry night sky, we can't help but wonder: Where did all this come from?

⚡ Looking at the *alternatives* to God can actually be a persuasive argument *for* the biblical God.

⚡ Eastern religions avoid the biblical God by saying that the universe has always been, that the universe is its own cause. This notion runs contrary to sound logic. Moreover, the 20th century has brought us undeniable evidence that the universe did have a beginning.

⚡ The steady state theory proposed that the universe eternally creates new matter out of nothing. But the theory was disproved by its own promoter when Fred Hoyle found that stars alone could not have produced the amount of helium observed in the universe today. This amount of helium *is* perfectly explained by a high density creation event. Also, nearby galaxies are not found in stages of early formation, as the steady state theory predicts.

⚡ The plasma theory proposes that electronically conducting gases produce a repelling effect between galaxies, causing the observed expansion. No version of the theory can account for the observed amounts of elements in the universe, or for the microwave background radiation, or for the decelerating rate of the universal expansion, or for the fact that the universe didn't run out of hydrogen a long time ago.

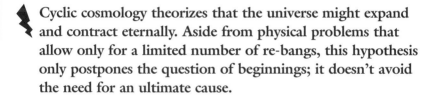 When the world's greatest minds (e.g., Einstein and Hawking) are put to the task of finding a way to propose a universe without a beginning, they find the task impossible, at least as far as can be shown by any theory that would line up with what we know of the real universe.

Cyclic cosmology theorizes that the universe might expand and contract eternally. Aside from physical problems that allow only for a limited number of re-bangs, this hypothesis only postpones the question of beginnings; it doesn't avoid the need for an ultimate cause.

Science is limited to what can be known by observation, which cannot be extended back before Planck time. But the limitations of science should not be confused with the limitations of reality. Science ends where the Bible begins.

Notes for Chapter 5

1. Stephen Hawking, quoted by Dennis Overbye, *Lonely Hearts of the Cosmos—The Story of the Scientific Quest for the Secret of the Universe* (New York: Harper Collins, 1991), p. 120.

2. Steven Weinberg, *Dreams of a Final Theory—The Search for the Fundamental Laws of Nature* (New York: Pantheon Books, 1992), p. 257.

3. John D. Barrow and Joseph Silk, *The Left Hand of Creation—The Origin and Evolution of the Expanding Universe* (New York: Basic Books, Inc., 1983), p. 13.

4. Ibid.

5. Ibid., p. 194.

6. Fred Hoyle, *The Intelligent Universe* (New York: Holt, Rinehard and Winston, 1983), p. 176.

7. Ibid., pp. 176-179.

8. P.J.E. Peebles, *Principles of Physical Cosmology* (Princeton, NJ: Princeton University Press, 1993), p. 570. Bracketed material added for clarity, replacing "at z ≥ 1," a reference to a great amount of redshifting, indicating great distance.

9. Eric Lerner, *The Big Bang Never Happened—A Startling Refutation of the Dominant Theory of the Origin of the Universe* (New York: Random House, 1991), pp. 217-218.

10. Ibid., pp. 300-301.

11. Ibid., p. 295.

12. Ibid.

13. Arthur S. Eddington, "The End of the World: From the Standpoint of Mathematical Physics," *Nature*, vol. 127 (1931), p. 450.

14. A. Vibert Douglas, "Forty Minutes With Einstein," *Journal of the Royal Astronomical Society of Canada*, vol. 50 (1956), p. 100.

15. Hoyle, p. 237.

16. Barry Parker, *Creation—the Story of the Origin and Evolution of the Universe* (New York & London: Plenum Press, 1988), p. 202.

17. Stephen W. Hawking, *A Brief History of Time—From the Big Bang to Black Holes* (New York and others: Bantam Books, 1988), p. 139.

18. Ibid., p. 136. Emphasis his.

19. Ibid., p. 139.

20. Carl Sagan, in "Introduction" to Stephen W. Hawking's *A Brief History of Time*, p. x.

21. Hawking, p. 139.

22. Letter from Stephen Hawking to the author, July 12, 1994.

23. Hawking, *A Brief History of Time*, pp. 136-137.

24. Alan Guth, interview with the author, June 8, 1994.

25. Albert Einstein, quoted by Nick Herbert, *Quantum Reality—Beyond the New Physics* (Garden City, New York: Anchor Press/Doubleday, 1985), p. 177.

26. Ron Cowen, "New Challenge to the Big Bang?" *Science News*, vol. 144 (October 9, 1993), p. 237.

27. Ibid., p. 236.

28. Stephen Hawking, quoted by Overbye, p. 103.

29. Barrow and Silk, pp. 71-72.

30. Richard Morris, *The Fate of the Universe* (New York: Playboy Press, 1982), pp. 136-137. Emphasis added.

31. George Smoot, interview with the author, May 6, 1994.

32. Howard J. Van Till, Davis A. Young, and Clarence Menninga, *Science Held Hostage* (Downers Grove, IL: InterVarsity Press, 1988), pp. 20-21.

33. Milton Rothman, "What Went Before?" *Free Inquiry*, vol. 13, no. 1 (Winter, 1992/93), p.12. Milton Rothman has served as professor of Physics at Trenton State College and as a research physicist at the Princeton Plasma Physics Laboratory. He authored *The Science Gap* (Prometheus Books, 1992).

34. Jeremiah Ostriker, interview with the author, July 13, 1994.

35. Parker, p. 10.

36. Carl Sagan, *Cosmos* (New York: Random House, 1980), p. 4.

37. Rothman, p. 12.

38. Victor J. Stenger, "The Face of Chaos," *Free Inquiry*, vol. 13, no. 1 (Winter, 1992/93), p.13. Victor Stenger is professor of Physics and Astronomy at the University of Hawaii. He wrote the book, *Not By Design: The Origin of the Universe* (Prometheus Books, 1988).

39. Arno Penzias, interview with the author, May 4, 1994.

40. Isaac Asimov, "What is Beyond the Universe?" *Science Digest*, vol. 69 (April, 1971), p. 69.

41. Parker, p. 195.

42. Fred Hoyle and Chandra Wickramasinghe, *Evolution from Space* (London: J.M. Dent & Sons, 1981), p. 144.

43. James Truran, interview with the author, May 6, 1994.

44. Albert Einstein, *Ideas and Opinions—The World As I See It* (New York: Bonanza Books, 1974), p. 375.

45. Paul Davies, *God and the New Physics* (New York: Simon and Schuster, 1983), p. 32.

46. John Mather, interview with the author, May 11, 1994.

47. George Smoot and Keay Davidson, *Wrinkles in Time* (New York: William Morrow and Company, 1993), p. 292.

48. Davies, p. 32.

49. Robert Jastrow, *God and the Astronomers,* second ed. (New York & London: W.W. Norton & Company, 1992), p. 105.

50. Ibid., pp. 106-107.

51. Carl Sagan, *Cosmos* (the television series), Episode 12.

52. Jastrow, p. 106.

53. Ibid., pp. 9-10.

54. Ibid., p. 103.

55. Parker, pp. 201-202.

56. Jastrow, p. 107.

The Horsehead Nebula, IC 434. —*Courtesy of Lick Observatory, University of California*

CHAPTER 6

Scientific Pointers to Creation

The heavens declare the glory of God;
the skies proclaim the work of His hands.
Day after day they pour forth speech;
night after night they display knowledge.
There is no speech or language
where their voice is not heard.
Their voice goes out into all the earth,
their words to the ends of the world.

—*Psalm 19:1-4*

A Skeptic's Questions:

So far you've been arguing philosophically or from negative scientific evidence. When you point out scientific problems for the non-God position, that's all negative evidence. Do you have any *positive* scientific evidence for an actual creation?

A Bible Believer's Response:

As a matter of fact, the consensus of the modern scientific community is that the universe had a beginning, a "creation event," as scientists often call it. Three broad lines of evidence make this conclusion practically inescapable: the laws of thermodynamics, Einstein's general theory of relativity, and the observations of astronomy.

Carl: Whoa! I'm having enough trouble with football fields and microwave ovens. Now you want to talk about thermodynamics and Einstein's general theory of relativity?

Fred: If you can throw around Christian terminology with your Christian friends, don't you think you could make the effort to get acquainted with some scientific terminology for your skeptical friends?

Carl: Well, I suppose it might come in handy sometime in a game of Trivial Pursuit.

Fred: Actually, what we're about to talk about makes just about everything *else* trivial by comparison. The familiar world of work and worry and family and friends can distract us from the big picture. Everything we experience is only possible because of a very big, supernatural event: the creation. By all rights, we shouldn't be here. Someone went to a great deal of trouble to make all this possible.

Carl: And you can show this with science?

Fred: That's where those scientific terms come in. When skeptics say, "Show me God," this is a good place to begin.

The Laws of Thermodynamics Point to a Creation Event

The First Law of Thermodynamics

Also called the law of conservation of mass and energy, this law states that matter and energy can be neither created nor destroyed. Matter and energy may each be *converted* to the other, as shown by Einstein's famous equivalence of mass and energy (summed up by the equation $E = mc^2$, where E is energy, m is mass, and c is the speed of light).

But neither mass nor energy can appear from nothing. Such an occurrence would be "a free lunch," as big bang theorist Alan Guth likes to say,[1] in contradiction to the common sense notion that there is no free lunch. And yet, there is no denying that the universe is here; so the universe itself appears to be a free lunch. But from the laws of physics we see operating today, creation is impossible as an ongoing event. That is, the conditions that we know hold true in our present universe prevent any possibility of matter springing out of nothing today.

Concerning the first law of thermodynamics, Isaac Asimov wrote: "This law is considered the most powerful and most fundamental generalization about the universe that scientists have ever been able to make."[2]

Thus the steady state theory's demand for a continual creation of new matter violates the first law of thermodynamics. To say that no new matter is being created is to agree with the Bible's statement that "the heavens and the earth were *finished*" (Genesis 2:1), that God "*rested*" from His work of creation (2:2).

The Second Law of Thermodynamics

The second law of thermodynamics tells us that the contents of our universe are becoming less ordered and more random. Left to themselves, things become disorganized. Things wear out. Ask any homeowner.

Even though the first law says that energy cannot be destroyed, the second law says that it does degrade, so that as energy radiates, less of it is available for mechanical work. Entropy (the amount of disorder in a system) never decreases in any physical interaction. And so the universe is wearing down.

Arthur Eddington showed that the energy of the universe must irreversibly flow from hot to cold bodies. Our sun is burning up billions of tons of hydrogen fuel every second. The earth's magnetic field is decaying; its rotation is slowing down. Stars, whole galaxies, are going to burn themselves out, never to light up again,* and matter will become more and more dispersed, less and less structured.

Thus we know that the universe cannot be eternal; it could not have been dissipating forever. If it had been eternally dissipating, it would have run down long ago beyond the point where we'd have stars shining. Working backwards, the law clearly points to a beginning. In fact, it points not only to a beginning, but to a highly *ordered* beginning. This raises the obvious question: If the universe is becoming less ordered, where did the initial order come from? Physicists have long been asking this question and have had no success in finding a natural solution.[3]

It is significant to note that two-and-a-half-thousand years before the birth of modern science, when the brightest thinkers were confident that the universe was unchangeable, according to all that they could observe, the Bible writers were in full agreement with this idea that the universe is "wearing out": ". . . the heavens are the work of your hands. They will perish, but you remain; they will all wear out like a garment. Like clothing you will change them and they will be discarded" (Psalm 102:25,26; also see Isaiah 34:4 and 51:6).

---

*The burning of hydrogen in stars is an irreversible process, resulting in heavier elements. These elements can never be converted back into hydrogen again. Thus the supply of hydrogen in the universe is growing smaller.

The ancients took their solemn oaths by heaven and earth, the most permanent and unshakable things they knew; the Bible warned its readers not to do so, for heaven and earth would be shaken, human intentions are frail, and only God is unshakable and eternal (Matthew 5:34-37, Isaiah 13:13, 54:10).

Only the Hebrews Got Their Cosmology Right

Part of the reason that the Bible writers had a higher view of God than of the universe was that they alone, among all the ancients, believed in a God who had created the universe— not in a magical, eternal Universe that gave birth to the gods. Only the Hebrews were true monotheists, and only they believed in a God who existed *before* the universe was created. Recognizing this, Fred Hoyle writes that "the general concept of gods located fairly and squarely within the Universe was common in ancient times throughout the Near East. The Hebrew departure from this position was evidently very great."[4]

To sum up: the second law of thermodynamics tells us that the universe must have had a beginning, and that it must have been a highly *ordered* beginning. The first law tells us that the universe could not have begun itself. From these two bedrock principles we can already infer that the universe was created by an entity from outside the universe and above natural law, i.e., by supernatural agency, one disposed to producing unfathomable order.

Einstein's General Theory of Relativity
Points to a Creation Event

To follow the equations of Einstein's general theory of relativity requires a level of mathematical knowledge that few of us possess. One story from the 1920s has it that a reporter asked physicist Arthur Eddington, a relativity expert, if it was true that only three people in the world understood Einstein's theory. After a long pause, Eddington finally replied, "I was just trying to think who the third person is."[5]

For our present purposes, there are just two main points to be made about general relativity: (1) all the testable predictions that it makes have been proven correct, and (2) the equations of general relativity imply that the universe cannot be static, but must be expanding or contracting. Even Isaac Newton knew that his laws of gravity implied that every star in the universe should attract every other star until eventually the entire universe converged. And even Newton had worked with four dimensions: three with spacial coordinates and one with a time coordinate. Einstein simply (well, not so simply) recognized the dependent relationship between the time coordinate and the other three.

Newton Lays the Groundwork for the 20th-Century Theory of a Creation Event

In his correspondence with Cambridge scholar, Reverend Richard Bentley, Isaac Newton pointed out the problems with a static universe, laying the groundwork for 20th-century cosmology. In 1692, Bentley wrote Newton that, according to his laws of gravity, all the stars of the universe must eventually pull each other together into a universal fireball, to which Newton responded:

. . . if the matter of our Sun & Planets and the matter of the Universe was evenly scattered throughout all the heavens, & every particle had an innate gravity towards all the rest & the whole space through which this matter was scattered was but finite: the matter on the outside of this space would by its gravity tend toward all the matter on the inside & by consequence fall down to the middle of the whole space & there compose one great spherical mass.[6]

Isaac Newton at age 49, in 1691.
—*Courtesy of the Mary Lea Shane Archives of the Lick Observatory*

Newton thus argued that the universe must be infinite and that all stars must be equally separated from one another. However, he soon realized that this solution was extremely unstable. The least imperfection in the equidistance between any stars would result in a chain reaction that would bring about the collapse of the entire universe. Newton, of course, never conceived of the possibility that the universe might be expanding, thus counteracting the gravitational force that must otherwise pull it together.

Newton had already shown that when an object was put in motion, its movement was relative to the movement of any observers. Consider the movement of a ball being tossed by passengers inside a horse-drawn coach. The ball may travel two feet relative to the passengers—while it travels *twenty* feet relative to an outsider watching the coach pass, because the coach's movement must be added to the movement of ball.

But Einstein showed that when an object was put in motion, the *time* it took to travel was also relative to the movement of any observers—as a result of the odd fact that light always travels at the same speed relative to all observers, no matter what their speed or direction. In other words, if you wanted to compute the speed that light travels from a flashlight being shined straight ahead from a coach, you would expect to have to add the coach's movement to the normal speed of the light. Instead, unlike a ball, light has the odd property of traveling at the same speed, both to the passengers and to those watching the coach pass.

The relativity of time becomes particularly noticeable at speeds near the speed of light: an astronaut traveling near the speed of light might age one day while the folks at home age one year (though the difficulty of reaching such speeds keeps such illustrations rather theoretical). Moreover, because of the relationship between mass and energy, to an observer on earth, the astronaut traveling at such a velocity would appear to gain weight and shrink in size (in the direction of motion) the nearer he approached the speed of light.

The description of such unusual phenomena accompanying bodies traveling at high velocities is known as the *special* theory of relativity, which Einstein published in 1905. This theory introduced the concept of space-time, showing how space and time are related to one another. *General* relativity (published in 1915) showed that properties of space-time precisely explain the force of gravity. Einstein reasoned that gravity could not actually be an attractive force that instantaneously acted on bodies at great distances, since special relativity said that nothing could travel faster than light. Rather, gravity is a consequence of the effect of mass on space-time. Great masses should noticeably "curve" the space around them—and "slow down" the time for any observer near them.

Thus we can picture the gravitational force of the sun, not as an attractive force that tugs on the planets, but as the mass of the sun curving the space around it, and so forcing each planet to follow a path that is as straight as it can go in curved space.

The prediction that massive objects should slow down the time for a nearby observer was conclusively proved in 1962 by the difference found between extremely accurate clocks running at the top and bottom of a water tower. Time ran slower for the clock nearer to the earth, where the earth's gravitational influence was greater. The results were in precise agreement with the predictions of general relativity.[7] Stephen Hawking describes how satellites now routinely depend on these predictions for their accurate

Mass Curving Space-Time

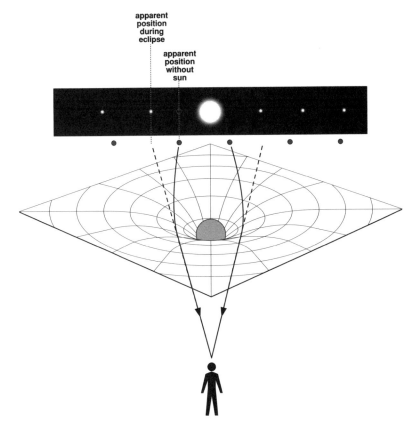

—*Illustration © 1995 Christopher Slye*

navigation systems: "If one ignored the predictions of general relativity, the position that one calculated would be wrong by several miles!"[8]

The correctness of general relativity theory was instantly implied by its accurate accounting for an anomaly in Mercury's orbit which Newton's laws could not explain. More evidence was forthcoming when two other predictions of the theory were observed: the bending of starlight when it passes the large mass of the sun and the change of light's frequency in gravitational fields. Sir Arthur Eddington first observed the bending of starlight passing the sun, though imprecisely, during his expedition to view the total solar eclipse of 1919 on the island of Príncipe (west Africa). Astronomer Walter Adams later observed the predicted wavelength shift in the white dwarf star, Sirius B. Today's measurements have shown the theory's predictions to be accurate to five places of the decimal, as accurate as measuring techniques allow.[9]

Einstein's equations show that, if there is enough mass in the universe, and if it is not too dispersed, the mass of the universe will actually cause all of space to curve in on itself until the universe becomes "closed," much like Newton's thought that all the bodies of the universe would fall together into a central mass. On the other hand, if the density of the universe is *lower* than a critical amount, then Einstein's equations implied that all the bodies of the universe should fly apart from one another. The universe should be expanding and decelerating with time.

Einstein decided that this description of an explosion had to be avoided at all costs. Either an expanding or a contracting universe could not be

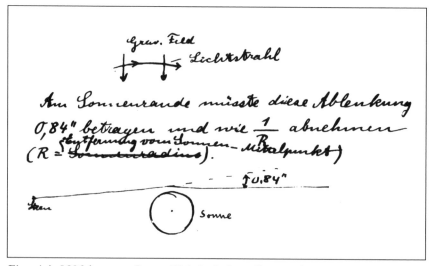

Einstein's 1913 letter to George Hale includes a diagram showing Einstein's prediction of exactly how Hale should find that a star's light path is bent by the sun.
—*Courtesy of The Huntington Library*

reconciled with the common scientific view that the universe was static and eternal. And so in a 1917 paper titled, "Cosmological Considerations on the General Theory of Relativity," he chose to distrust the logic that had led him to his equations and to add his famous "cosmological constant." This odd force was required to compensate for gravity and to increase in strength with distance, unlike any other force known to physics. And it was necessary that it be applied in a very precise degree in order to keep the universe static: perfectly balanced between a universe that begins to expand or one that begins to contract into a "big crunch."

In 1922 Russian mathematician Alexander Friedmann found an algebraic error in Einstein's proofs for a static universe. By correcting the mistake and by discarding Eintstein's cosmological constant, Friedmann discovered that Einstein's "static" universe was virtually impossible, and that the universe must be either open or closed, i.e., expanding or contracting.

Working independently a few years before, Dutch astronomer Willem de Sitter also found solutions to Einstein's equations that required the universe to be *expanding*—even with the cosmological constant. And British astronomer Arthur Eddington independently found that, even with the cosmological constant, the equilibrium Einstein had achieved was an "unstable" equilibrium—a slight shove either way would send the universe into expansion or collapse.

Einstein later chided himself for introducing his famous fudge factor in order to make his theory fit. He called the addition of his cosmological constant "the greatest blunder of my life."[10] He wrote: "The mathematician Friedmann found a way out of the dilemma. His results then found a surprising confirmation by Hubble's discovery of the expansion [of the universe]."[11] After this Einstein wrote not only of the necessity for a beginning, but of his desire "to know how God created this world. I am not interested in this or that phenomenon, in the spectrum of this or that element. I want to know His thought, the rest are details."[12]

And so, especially since the early 1960s, when it became possible to precisely measure the predicted effects of general relativity in test after test, science has turned from its disposition to hold to an eternal universe, holding instead to a universe with a beginning, as the general theory of relativity clearly predicts.

Astronomical Observations Point to a Creation Event

As described in the next chapter, observational evidence from astronomy also clearly indicates that the universe is expanding. The scientific consensus is that the expansion points back inescapably to a time when all the bodies in the universe had a common origin in time and space. Such an origin is impossible without an Originator. As we will see, there has never been a time in history when humankind has had such an unmistakable message from space to tell us that there is a Supernatural Creator. And the message doesn't stop there.

Shocking Summary Statements
and
Stimulating Conversation Starters

❧ Positive evidence for an actual creation event comes to us from three broad lines of scientific evidence: the laws of thermodynamics, Einstein's general theory of relativity, and the observations of astronomy.

❧ The first law of thermodynamics states that matter and energy can be neither created nor destroyed. To say that no new matter is being created today (in contradiction to the steady state theory) is to agree with the Bible's statement that "the heavens and the earth were finished," that God "rested" from His work of creation.

❧ The second law of thermodynamics tells us that the universe is becoming less ordered. It's "wearing out," just as the Bible says "the heavens . . . will all wear out like a garment." But the universe could not have been dissipating from eternity past, or we'd have no stars shining today. According to this second law, the universe must have had a beginning, and *it must have been a highly ordered beginning.*

❧ Einstein's general theory of relativity tells us that, depending upon its density, the universe must be either collapsing or expanding. Einstein was forced to renounce his belief in an eternal universe and admit that the universe must have had a beginning.

❧ Among all the ancient peoples, only the Hebrews got their cosmology right. Only they believed in an eternal, omnipotent God who gave the universe its beginning, not in a magical, eternal universe that gave birth to the gods.

Notes for Chapter 6

1. Heinz R. Pagels, *Perfect Symmetry* (New York: Simon and Schuster, 1985), p. 316.

2. Isaac Asimov, "In the Game of Energy and Thermodynamics You Can't Even Break Even," *Journal of the Smithsonian Institute* (June, 1970), p. 6.

3. Paul Davies, *God and the New Physics* (New York: Simon and Schuster, 1983), p. 168.

4. Fred Hoyle, *The Intelligent Universe* (New York: Holt, Rinehart and Winston, 1983), pp. 236-237.

5. Stephen W. Hawking, *A Brief History of Time* (New York: Bantam Books, 1988), p. 83.

6. Robert Newton, in his letter to Reverend Richard Bentley, dated December 10, 1692, cited by George Smoot and Keay Davidson, *Wrinkles in Time* (New York: William Morrow and Company, 1993), p. 26.

7. Hawking, p. 32.

8. Ibid., p. 33.

9. Hugh Ross, *Big Bang Ripples: Proof Positive for the Creation Event* (audio tape of an announcement by Dr. Ross, produced in Pasadena, CA, by Reasons to Believe, 1992).

10. Albert Einstein, cited by Richard Morris, *The Fate of the Universe* (New York: Playboy Press, 1982), p. 28.

11. Albert Einstein, cited by Barry Parker, *Creation—The Story of the Origin and Evolution of the Universe* (New York & London: Plenum Press, 1988), pp. 53-54.

12. Albert Einstein, cited by Nick Herbert, *Quantum Reality—Beyond the New Physics* (Garden City, New York: Anchor Press/Doubleday, 1985), p. 177.

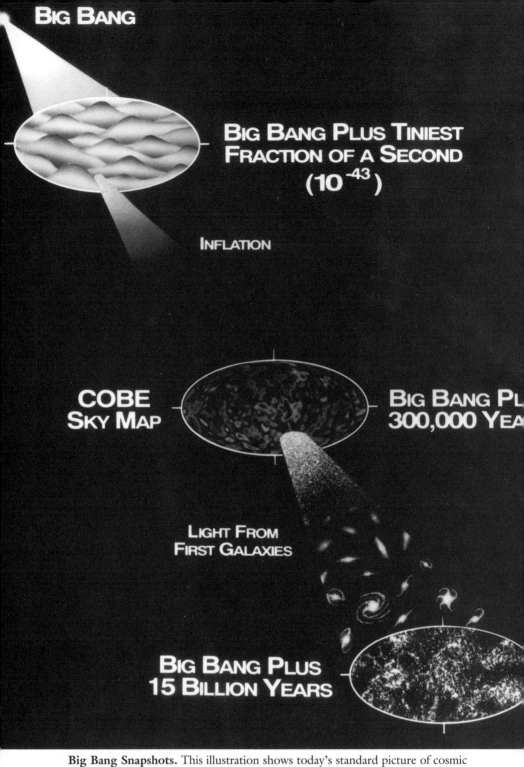

BIG BANG

BIG BANG PLUS TINIEST FRACTION OF A SECOND
$$(10^{-43})$$

INFLATION

COBE SKY MAP

BIG BANG PL 300,000 YEA

LIGHT FROM FIRST GALAXIES

BIG BANG PLUS 15 BILLION YEARS

Big Bang Snapshots. This illustration shows today's standard picture of cosmic history. But how do we know it's true? The COBE satellite has brought us data from a time before galaxies formed, far earlier than anything radio or optical telescopes can show us. This chapter is devoted to exploring this and other observational evidence. —*Courtesy of NASA*

CHAPTER 7

The Big Bang theory

*There is no doubt that a parallel exists
between the big bang as an event and the
Christian notion of creation from nothing.*

—George Smoot

A Skeptic's Questions:

Here's one place where your Bible definitely conflicts with science. I know you Christians don't believe in the big bang theory or that the universe has been around for billions of years.

A Bible Believer's Response:

Some Christians believe in the big bang theory and some don't. But everyone who believes in the big bang theory must be a creationist, of sorts. Modern science's most favored theory of beginnings, the big bang theory, describes a creation event that defies atheism and pantheism, but harmonizes with the Bible.

Carl: Don't tell me you believe in the big bang theory!

Fred: Why? What's your opinion of the big bang theory?

Carl: Well, it's completely worthless.

Fred: I know, but I'd like to hear it anyway.

Carl: Now stop that! All right, you got me again. But the Bible says nothing about a big bang. The whole idea is crazy.

Fred: The Bible says nothing about the atomic table of elements. Is all modern chemistry crazy? You shouldn't feel threatened by it. The big bang theory by itself doesn't explain anything about ultimate origins. It only raises questions that science can't answer. What caused the big bang? How did all the right ingredients for the big bang get there? As we've seen, science hasn't been able to tell us anything about that. As astronomer Robert Jastrow puts it:

> There is no explanation in the Big Bang theory for the seemingly fortuitous fact that the density of matter has just the right value for the evolution of a benign, life-supporting Universe.[1]

Carl: Well, the big bang theory is just that: it's a theory. It doesn't prove anything.

Fred: As a matter of fact, we're about to talk about what astronomers are actually seeing through their telescopes. This is where the observational evidence comes in to back up all the theories.

The Basis for the Big Bang Theory

George Smoot, leader of the COBE satellite team that first detected the cosmic "seeds" of the early universe, writes: "Until the late 1910's, humans were as ignorant of cosmic origins as they had ever been. Those who didn't take Genesis literally had no reason to believe there had been a beginning."[2] Perhaps the best way to follow the logic behind the big bang theory is to follow the process, step by step, by which the discoveries were made that led to it.

Red Shifts

As early as 1914, American astronomer Vesto Slipher announced that almost all of the nebulae (as the galaxies were then called) he had measured were receding from us at high velocities. The measurements were a result of spectroscopy, the technique of using a prism to separate light into its component colors, usually used to determine the elements that compose a

Vesto M. Slipher (1875-1969) about 1905. —*Courtesy AIP Emilio Segré Visual Archives*

star. Slipher had set up a spectroscope for Percival Lowell's observatory in Flagstaff, Arizona in 1901, and so was surprised in 1912 to find how much the nebulae were shifted to the red side of the spectrum, compared with the stars. The characteristic bands of hydrogen, helium, etc., in the nebulae were clearly identifiable, but they were moved from their normal position far to the red side.

By this time, red and blueshifting were already well understood by astronomers to be an expected effect of stars moving relative to us, since the Doppler effect demands a shift in frequency with light waves as well as sound waves (as when a car horn rises in pitch as it approaches us and falls again when it travels away). When a light source travels away from an observer, the light waves are stretched, causing longer wavelengths and making the light appear redder. Light waves from a source moving *toward* an observer "bunch up," causing shorter wavelengths and a bluer color. What surprised Slipher was the extremely high degree of redshifting he observed in the nebulae compared to the moderate amounts he routinely measured in stars.

The tiny, faint spiral and elliptical nebulae were thought to be gas or dust clouds in the same vicinity as the stars—all within our Milky Way. Stars were sometimes red and sometimes blueshifted, apparently indiscriminately. In contrast, almost all the nebulae were *greatly* redshifted (receding from us). The few that were blueshifted (approaching us) were in the same small group. All the nebulae were receding at an extremely high velocity, some as high as a thousand miles per second. Slipher didn't know quite what to make of this, since he firmly believed that all the stars and nebulae were simply drifting about randomly through the universe.

The consensus of the day was that the universe was static and eternal. It was also assumed that our Milky Way Galaxy *was* the universe, and so the movement of these stars both toward us and away from us was seen as the indiscriminate or swirling movement of the entire universe. The thought that the faint nebulae might be distant "island universes" of their own and not part of our galaxy, as proposed by Immanuel Kant back in the eighteenth century, was not taken seriously.

Einstein was acquainted with astronomical observations showing that stars were apparently drifting in no particular direction. When he realized that his new general relativity theory implied that the universe was expanding, he thought that something must be missing from his equations, and so added his cosmological constant to keep the universe static.

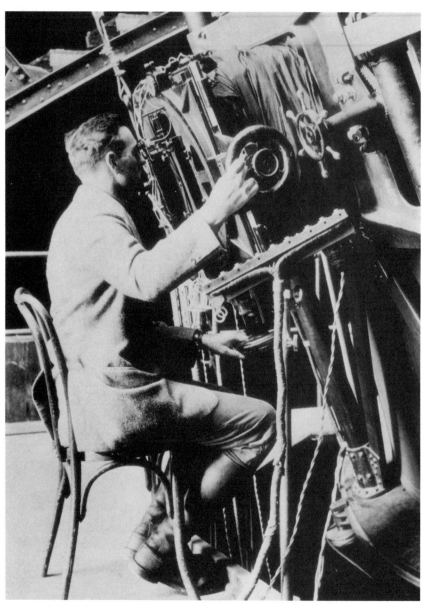

Edwin Hubble takes his customary viewing position in 1924, where he spent most of the 1920s: at the Newtonian focus of Mount Wilson's 100″ Hooker telescope.
—*Courtesy of The Huntington Library*

Hubble Discovers Galaxies

The first step toward putting the facts together came with Edwin Hubble's discovery that the nebulae were not just nearby gas clouds, but were themselves galaxies of stars like our own Milky Way. After fighting his way through a decade of boxing, dueling, and World War I combat, this Rhodes scholar decided to devote the rest of his life to what he considered to be the most exciting pursuit ever: astronomy. Hubble saw this, not as a career move, but as a "calling." He quit his newly established, successful law practice and, after receiving his Ph.D. in astronomy, spent the 1920s away from the roar, patiently peering into a 100-inch reflector telescope at California's Mount Wilson Observatory (Slipher had used a 24-inch refractor).

Using the world's most powerful telescope, Hubble scrutinized the nebulae, photographing their outer regions in long exposures—sometimes extending over several nights—until at last he was able to discern stars. Having thus established that these nebulae were actually star systems, he soon offered additional proofs that they were at great distances from us compared to the stars and globular clusters of our Milky Way.

The 100" Hooker Telescope at Mount Wilson, CA. Edwin Hubble used this rivet-mounted telescope, the largest in the world for many years, to make his discovery of an expanding universe. Using the known traits of Cepheid stars, he was able to measure the distances of distant galaxies. Combining these distances with knowledge of their redshifts, he formulated the Hubble law: the farther the galaxy, the faster it recedes from us. —*Courtesy of the Carnegie Institution*

Edwin Hubble at his desk at Mount Wilson in 1935. —*Courtesy of The Huntington Library*

How Far Is a Star?

How does one measure distances to the stars? The most accurate method is by trigonometric parallax, performed by measuring the position of a nearby star from two points on opposite sides of the earth's orbit. Distance can be calculated by observing the small angular displacement between these two measurements. Astronomers were amazed to find out how far the nearest stars are. In fact, only a tiny fraction of the 100 billion stars of our galaxy is within reach by this measurement—the parallax method becomes virtually useless for stars that are more than about 500 light-years distant.

To find the distances to the nebulae, Hubble at first used the known magnitudes (brightness levels) of certain types of nearby celestial objects as standards to measure against those he found in the much fainter systems outside our galaxy. A useful law of optics tells us that the brightness of a light will vary inversely with the square of its distance. That is, if we know that two stars have the same intrinsic brightness, and one of them is four times fainter than the other, then we know that it is twice as far away.

Hubble made use of his knowledge of the brightness of nearby novas (stars that suddenly flare up intensely and then fade) and star clusters, but his most accurate measurements made use of a particular type of star known as the Cepheid variable. In 1912 Harvard's Henrietta Leavitt had shown that this class of star undergoes regular periods of brightening, and that the length of this cycle correlates closely to its true brightness. The longer the periods of time between peak brightness, the brighter the Cepheid stars. Thus the pulsation cycles and brightness of Cepheid stars in a nearby system called the small Magellanic cloud provided the tools to measure the actual brightness of any Cepheids that could be found in distant galaxies, regardless of their apparent brightness. Once the actual brightness was known, the distance could be calculated by use of the inverse square law.

Hubble searched for and found Cepheid stars in other galaxies, allowing him to make the first knowledgeable estimates of their distances. Scrutinizing these galaxies, he also discovered numerous star clusters and bright novas. He used measurements of their brightness (compared with the known average brightness of closer novas and star clusters) to supplement and confirm the Cepheid measurements. In 1924 he found the distance to the Andromeda nebula (the galaxy nearest to our Milky Way) to be 900,000 light-years (about half the distance shown by improved measurements today).

To find the distances of farther nebulae, he developed two more measuring criteria. Beyond 1,000,000 light-years, he could no longer find Cepheids, but he could use the brightest stars in the nebulae with Cepheids to compare with the brightest stars in farther galaxies. Beyond 6,000,000 light-years, he used the total luminosity's of the entire nebulae as his measuring tool. This gave his telescope a measuring range of up to 250,000,000 light-years.

The Universe Is Exploding!

By 1929, Hubble had determined the distances of 24 nebulae outside our galaxy for which degrees of redshifting were also known, mainly from Slipher's work. He found that though the nearby stars are not retreating from us, the distant galaxies are. Moreover, the farther galaxies are retreating at a faster rate than the nearer galaxies. In fact, he found that there was a precise, linear relationship between the distance and velocity of the galaxies. A galaxy that is twice as far as another is traveling twice as fast. His publishing of these results upset the world of science and inspired more intensive studies in cosmology.

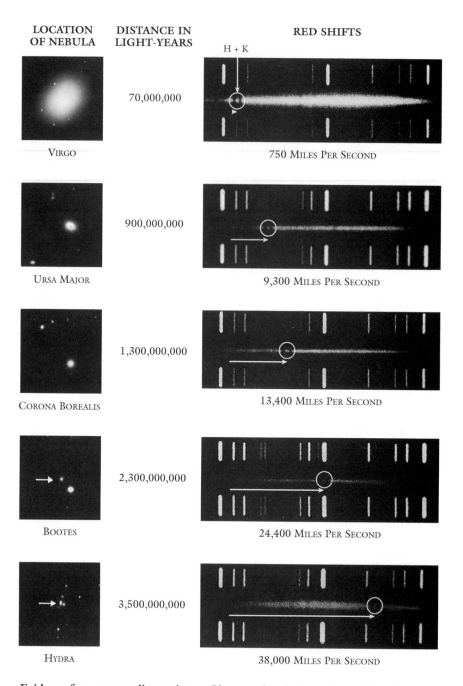

LOCATION OF NEBULA	DISTANCE IN LIGHT-YEARS	RED SHIFTS
VIRGO	70,000,000	750 MILES PER SECOND
URSA MAJOR	900,000,000	9,300 MILES PER SECOND
CORONA BOREALIS	1,300,000,000	13,400 MILES PER SECOND
BOOTES	2,300,000,000	24,400 MILES PER SECOND
HYDRA	3,500,000,000	38,000 MILES PER SECOND

Evidence for an expanding universe. Photographs of galaxies in the left column correspond to their spectra in the right column. Galaxy photos at left show their decreasing size and apparent brightness with distance. Their corresponding spectra at right show increasing redshifting. Notice, as you scan each spectrum from top to bottom, the increasing displacement of the encircled double lines (the H and K lines of calcium). These redshifts resulted in Hubble's velocity/distance relation and evidence for an expanding universe. —*Courtesy of Robert Jastrow, Mount Wilson Institute*

While debates raged over his findings, Hubble pressed on to perfect his methods; and along with spectroscopic specialist M.L. Humason, he provided clearer evidence of the distance-velocity correlation from numerous measurements of the galaxies, finding that the farthest galaxies he could measure (at 250,000 light years) were receding at 26,000 miles per second, nearly one-seventh the speed of light. By 1936, Hubble had probed the universe to the limits of his 100-inch telescope.

Growing evidence from other astronomers during this time, along with more recent evidence from more powerful telescopes and new methods, has served to show that Hubble had underestimated distances to the galaxies; but the new evidence has also served to confirm the distance-velocity relationship. By noting the higher velocities of the more distant galaxies (distant in time as well as space, since we are observing them as they were in the distant past), astronomers learned that the expansion is not merely a matter of galaxies retreating at an even pace. In earlier periods of our universe, the galaxies were traveling away from each other at much higher velocities; the universal expansion is now decelerating from a powerful, initial surge. Thus the consensus of modern science is not simply that the universe is expanding—it is exploding!

This description of an explosion—with the universe simultaneously expanding and decelerating (as opposed to an evenly paced expansion)—is precisely what Einstein's field equations had predicted. In 1917, Dutch astronomer Willem de Sitter had found a solution to Einstein's general relativity equations that also implied this velocity-distance relation—though de Sitter himself did not describe his results this way. His mathematical model was intended to show a static universe without time— and without matter in it— and it predicted that light

Edwin Hubble views through the 48-inch Schmidt telescope on Palomar Mountain (California) in 1950. —*Courtesy of The Huntington Library*

Hubble's velocity vs. distance plot of 1929.
The black discs and full line represent the Hubble law fitted to the nebulae individually. The circles and broken line represent the solution combining the nebulae into groups.

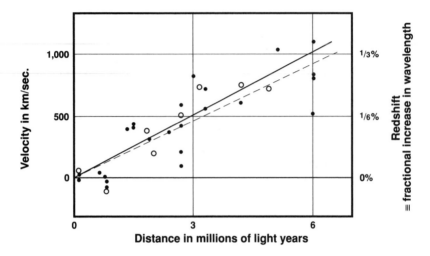

Hubble and Humason's 1931 velocity vs. distance diagram

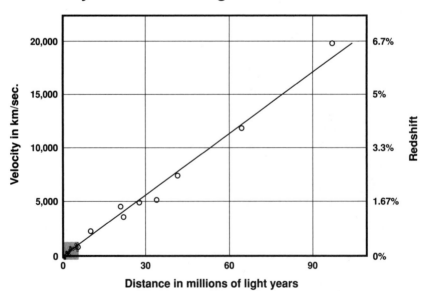

Notice the clarity of the linear relationship between velocity and distance when Hubble extended his observations to farther galaxies, as in the lower graph. The shaded portion of the lower graph represents the small amount of data Hubble had to work with from his earlier observations. —*Illustration © 1995 Christopher Slye*

from distant objects should be redshifted from our perspective. Though de Sitter himself did not recognize that this implied expansion, the expected expansion of the universe became known as the "de Sitter effect." When matter was added to de Sitter's equations, it was shown that any two masses would separate from one another—a universe filled with objects would expand.

How Old and How Big Is the Universe?

NOTE FROM CARL: I RECOMMEND THAT ANYONE WHO ISN'T REALLY INTO THE TECHNICAL STUFF SKIPS THIS SECTION.

Hubble's 1929 discovery, now known as the Hubble Law, tells us that all distant galaxies are retreating from us with a velocity that is directly proportional to their distances from us. In other words, if one galaxy is twice as far from our Milky Way as another, we will find that the galaxy that is twice as far is moving away from us twice as fast. Robert Jastrow (founder of NASA's Goddard Institute and now head of the Mount Wilson observatory, where Hubble made all his discoveries) writes: "The Hubble Law is one of the great discoveries in science: it is one of the main supports of the scientific story of Genesis."[3] And Jastrow, we should bear in mind, is a self-proclaimed agnostic (according to his writings as well as his recent interviews with me).[4]

Cosmologists express the precise velocity/distance relationship with a number called the Hubble constant (usually written H_0). The value of H_0 is critical because, if we could be certain of this number, we could determine the size and age of the universe. Efforts to establish the Hubble constant have not resulted in universal agreement, and thus cosmologists presently assign the universe a broad age range between 8 and 20 billion years (meaning also that the universe spans between 8 and 20 billion light-years across).

Astronomers express the Hubble constant in terms of kilometers per second (velocity) per megaparsec (distance). A parsec is the distance of an object from earth when the object varies one second of an arc when viewed on opposite sides of the earth's orbit (that is, viewing the object at times six months apart). The parsec (meaning parallax of one second) equals about 3.26 light-years; a megaparsec is a million times this amount. Recent calculations for the Hubble constant range between about 50 and 90 kilometers per second per megaparsec. A Hubble constant at the low end of this range results in an older and larger universe than a Hubble constant at the high end, because slower moving galaxies would take more time to reach their present distances.

How do astronomers calculate the Hubble constant? All distance calculations start with the Cepheid variable stars, the same reference Edwin Hubble used and still the most reliable one. Cepheids are observable up to about 30 million light-years away. Today, other "standard candles" are also employed, each with its own method for determining a light source's absolute magnitude (that is, its actual brightness, no matter how faint it may appear because of distance or intervening dust).

These standard candles include RR Lyrae stars (old yellow variable stars, observable to about 10 million light-years), planetary nebulae (rings of gas thrown off of dying stars, observable to about 75 million light-years), and spiral galaxies (which can use the Tully-Fisher method up to about 100 million light-years). To apply the Tully-Fisher method to a spiral galaxy, astronomers first make radio observations to determine the galaxy's rate of rotation. Knowing that a faster rotation rate means that the galaxy has more mass (according to Newton's laws), astronomers can calculate the galaxy's absolute brightness. As in all methods, the absolute brightness is then compared to its apparent brightness in order to calculate its distance.

The close agreement between each of these methods have helped to confirm the others. Each yield a Hubble constant that is at the high end of the range, indicating a universe that is relatively young and small. However, astronomers have another standard candle that is not in agreement with the others: the type Ia supernovas. These supernovas briefly shine as brightly as their host galaxy and are observable up to 100 million light-years away. Astronomers like Allan Sandage (considered Hubble's heir) feel they understand this particular type of supernova very well, having amassed considerable observational data that make the type easily identifiable by its spectrum, light curve, and peak absolute magnitude.

Cosmologists are thus put in the position of having to choose sides. Those who trust the Ia supernovas for their calculations believe the universe is 15 to 20 billion years old. Those who trust the other standard candles believe the universe is closer to 8 to 12 billion years old. Very recently, however, discoveries of the variations in Type Ia supernovas have tended to bring this standard candle almost in line with the others, meaning that the universe may be on the younger side of our range. Today the repaired Hubble Space Telescope is trained on Cepheid variable stars in the Virgo Cluster galaxies in order to resolve the dispute. New observations of Cepheids, which are needed to better calibrate all the other methods, may help settle the issue in the very near future.

Hubble Trouble. The Hubble Repair Mission turned the Hubble Space Telescope from an embarrassing flop into the world's farthest reaching optical telescope, capable of helping us begin to answer many questions about the early universe. —*NASA*

The Big Bang

In 1922 Russian mathematician Alexander Friedmann discovered (like de Sitter before him) that Einstein's "static" universe was virtually impossible, based on the general relativity equations, and that the universe must be either expanding or contracting. In 1927, Belgian physicist Georges Lemaître came to the same conclusions, though he had never heard of Friedmann's work. By 1933 Lemaître put the facts together in a way that has since earned him the title, "father of

Alexander Friedmann (1888-1925) in Moscow, 1925.
—*Courtesy of Leningrad Physics Technical Institute (and AIP)*

the big bang theory." The words "big bang" were actually coined much later as a disparaging term by astronomer Fred Hoyle, since, as he says, "The big bang theory requires a recent origin of the Universe that openly invites the concept of creation,"[5] a concept Hoyle felt that science could not take seriously.

Lemaître called his beginning point a "primeval atom." Today it is more accurately called a "singularity," which may be defined as the whole of space shrunk down to a size of zero volume. This odd state was at first studiously avoided by the physicists who examined the theory. Acting on Einstein's suggestion in 1933, Lemaître tried to find a way to apply the equations with different results. He tested models in which the universe did not expand at the same rate in every direction, but these merely brought him back to a singularity more quickly (more recently in the past).

Are We in the Center?

To understand why all the galaxies should appear to be flying away from *us*, imagine that you drew a number of dots on a balloon with a marker. Of course, no dot would be in more of a central position than any other. Then as you blew up the balloon, from the viewpoint of each dot, all the other dots would retreat from it. This also explains why Hubble found that the velocity of each galaxy's movement away from us is directly proportional to its distance from us. If we make one dot our point of reference, then a dot that is twice as far from it as another will move at twice the velocity from it.

This means that, as we turn the clock of our universe backward, at some point in the past, all galaxies must reach us at the same time: a galaxy that is twice as far from us as another will reach us at the same time because it was moving at twice the velocity. This precise, linear relationship of distance to velocity (acknowledged by all modern astronomers) requires that all galaxies had a common starting point at some time in the past.

Though some may like to argue that we can't know that such a universal convergence really resulted in a singularity, since all known laws of physics break down under the incredible pressures involved in squeezing the universe down into such a point, the fact is that the laws we *do* know don't break down until the density reaches 10^{96} times that of water, when the universe was 10^{-43} seconds old (Planck time).* As astronomers John Barrow and Joseph Silk point out, "such a fantastic density is already pretty singular by anybody's standards."[6]

---

*10^{96} is shorthand for 1 followed by 96 zeros. 10^{-43} is shorthand for a fraction of a second that would be written as 1 preceded by 42 zeros, preceded by a decimal point.

Microwave Background—Remnant of the Big Bang?

In 1927, George Lemaître, who first predicted the early dense state, predicted that his "primeval atom" might still be detected in the form of remnant radiation. Early big bang theorists George Gamow, Ralph Alpher and Robert Herman wrote that the heat radiating from the primeval explosion must still exist, since, unlike the heat from any other heat source, there is nowhere to which this primeval heat can escape. The heat from an oven escapes into the surrounding air; the heat from a volcano escapes into the atmosphere; but there is nowhere "outside" the universe to which the big bang radiation can escape.

Alpher and Herman even calculated in 1948 that this radiation should now be expected to be present everywhere in space at a temperature of about 5 degrees Kelvin (above absolute zero, the lowest possible temperature). They talked to several experimentalists about looking for this big bang left-over radiation, but could find no takers.

In 1965, without looking for it, two physicists at AT&T Bell Laboratories in New Jersey found it. At first, Bell's Arno Penzias and Robert Wilson were perturbed because, while trying to refine the world's most sensitive radio receiving device (a large horn-shaped antenna cooled down to nearly absolute zero so that it would be sensitive to extremely low-temperature radiation), they couldn't eliminate an unknown source of noise that corresponded to a temperature of about 3 degrees Kelvin.

Other researchers had used the antenna to track the Echo I and Telstar communication satellites. They had ignored the small level of background hiss and had simply reset their zeros. But Penzias and Wilson wanted to be sure they could accurately detect microwave radiation from the Milky Way Galaxy. Moreover, they were just plain curious. No matter where they pointed the receiver in the sky, this level of radiation remained constant.

At first they attributed it to bird droppings in their antenna. After removing the offending matter and eliminating every other conceivable fault over a period of months, they began to realize that the source of their problem was not their equipment—in fact the source of the radiation had to be beyond the earth's atmosphere, beyond the solar system, even beyond our galaxy and the other visible galaxies.

If the source had been influenced in any way by our atmosphere, the reading would have varied as they pointed the detector in different directions. Radiation entering our atmosphere would have a lot more atmosphere to penetrate when the detector was pointed toward the horizon than when it was pointed straight up—but the hum remained the same no matter where they pointed the detector. If the source had been our own Milky Way Galaxy or any other particular body in the heavens, the noise would have varied as the earth rotated—but the noise remained constant day and night and at different seasons later in the year as the earth revolved around the sun.

At about the same time, several physicists at Princeton University, led by Robert Dicke, determined that the radiation left over from the early hot universe should be able to be detected, no longer as light, but after stretching along with the expanding universe (that is, after redshifting through the spectra of visible light and radio wavelengths during its journey since the beginning of time) it should now be detected in the form of microwaves. Actually, a number of physicists had by this time predicted it. Robert Dicke persuaded two experimenters to look for the radiation and actually started construction on a small horn-like radio telescope on top of a campus building.

When the Princeton group heard of the Bell Lab discovery, they realized they'd been scooped. After a telephone call between the Bell and Princeton groups, Bell's Wilson was still hesitant to believe they had discovered anything having to do with the big bang. He had been a student of Fred Hoyle, and thus was much more disposed to believing in the steady state theory at the time. Each team wrote a paper which appeared in the same issue of *The Astrophysical Journal*; but the Nobel prize went to Penzias and Wilson in 1978, after many others had substantiated and refined the measurements.

Robert Wilson (left) and Arno Penzias stand in front of the Holmdel, New Jersey, antenna with which they unexpectedly detected the microwave background radiation in 1965. —*Courtesy of AT&T Bell Laboratories*

Ma Bell Gets a Message from Space

Interview with Arno Penzias[7]

Heeren: Could you tell me something about what you and Robert Wilson were looking for in 1965 and what you actually found?

Penzias: We were attempting to measure the high latitude radiation of the Milky Way at about 22 centimeters. In order to make sure our equipment was actually working, we first wanted to get a zero by measuring at a higher frequency, a shorter wavelength—7 centimeters—at which we expected no galactic component whatsoever. And if we got a zero there, then our other measurement would at least be meaningful. So we were attempting to make sure that we could in fact measure the absence of radiation from the Milky Way, when we in fact found radiation, which was coming, evidently, from beyond the Milky Way. And what we found was radiation for which there is no known source in the universe.

Heeren: I understand that Robert Wilson had been a student of Fred Hoyle and was probably disposed to believing in a steady state theory before 1965.

Penzias: He, like most physicists, would rather attempt to describe the universe in ways which require no explanation; there's the economy of physics. And since science can't *explain* anything—it can only *describe* things—that's perfectly sensible. If every time you wanted to describe a new phenomenon and you found an explanation for it, you'd be in a lot of trouble. Because you'd say the tree gets green because it *wants* to. Or it gets green because nature wanted it that way. Or a fairy comes there every March, or something like that. Each one of those are explanations. Whereas in fact all you describe, you describe capillary action which is the way molecules behave based on some deeper description.

Now, on the other hand, if you have a universe which has always been there, you don't explain it, right? Somebody asks you, "How come all the secretaries in your company are women?," you can say, "Well, it's always been that way." That's a way of not having to explain it. So in the same way, theories which don't require explanation tend to be the ones accepted by science, which is perfectly respectable and the best way to make science work. And science works in all cases except those issues where description is inadequate.

Heeren: That's the way you can keep the theories within the realm of science: by making sure you're only talking about descriptions.

Penzias: Right.

Heeren: But then having said all that, then what happened to his view and to yours to change them?

Penzias: Well, the steady state theory turned out to be so ugly that people dismissed it. The easiest way to fit the observations with the least number of parameters was one in which the universe was created out of nothing, in an instant, and continues to expand. Physicists normally would like a model in which there are no external parameters. So what we find—the simplest theory—the one that the astronomers normally espouse, is a creation out of nothing, the appearance out of nothing of a universe.

What does Robert Wilson say about all this? How long did it take him to acknowledge that the background radiation was actually a remnant of the big bang? And was the cosmic background radiation actually attributed first to bird droppings?

Interview with Robert Wilson[8]

Heeren: Now in some of the books I've read it actually talks about you folks going out there and scraping off some bird droppings, because you thought, "Well, it must be something that's hanging in there no matter where we point it in the sky." Is that true?

Wilson: We looked at several things relating to the telescope itself. And cleaning up after the birds was one of them. In fact that did knock a little bit off our measured temperature. There was some radiation from the bird droppings.

Heeren: I understand that you had been a student of Fred Hoyle and were probably disposed, as so many people were then, to believing in a steady state theory before 1965. Is that true?

Wilson: Well, I really didn't actually take a course from Fred Hoyle.

Fred was around at Caltech. I guess I was a student in the sense that I sat in on a course in cosmology that he taught. But yes, that connection was there. That was true, that I philosophically liked the steady state. And clearly I've had to give that up.

Heeren: What kind of period of time was involved in giving that up?

Wilson: Oh, this whole thing was spread out a lot, because we had the original problem, so to speak. And then along came this wacky idea from Princeton. They were actually thinking about multiple big bangs. Arnold and I were very happy to have some explanation of what was going on, but neither one of us, I think, bought the exact cosmology right away. So it was a matter of at least some months and probably more like a year.

Heeren: In general, some people feel that evidence for a beginning for the universe also provides evidence for some kind of—something outside of nature to have created nature. It seems that if you compare the steady state with the big bang, there's *one* theory that has the universe that's here eternally—and so maybe there's no need for anything else—then you have the *other* theory that says that everything started at one point, which would then require that something came out of nothing. And that makes me wonder if this then would be indicative of a Creator.

Wilson: Certainly there was something that set it all off. Certainly, if you are religious, I can't think of a better theory of the origin of the universe to match with Genesis. It may be the creation stories are a bit anthropomorphic: you know, people are created and live and die, and so the creation stories probably follow the same general pattern.

Heeren: But that is a very big difference between Genesis and the other creation stories. I'm probably a better student of history than I am of science. If you go back into comparative religions, in primitive religions, you find that the Hebrews alone had a concept of a creation event, whereas all the other religions seemed to have this amorphous blob that always was, this watery mass that everything then came out of, including the gods, after that. So there's a very big difference there.

Wilson: There is a big difference there, isn't there? Well, it [the big bang] certainly fits with that.

According to the basic framework of standard cosmology, temperatures in the early universe were so high that atoms could not form. Approximately 300,000 years after the creation event, when temperatures had dropped to about 3,000 degrees, matter finally decoupled from radiation, and the universe turned from opaque to transparent. Radiation was thus released in all directions, streaming through the history of the universe and reaching us today in the form of the cosmic microwave background radiation.

For readers who have been waiting for an explanation of this mouthful, the term *cosmic microwave background radiation* breaks down as follows: "Cosmic" means it has the entire universe as its source. "Microwave radiation" is defined as radio waves with wavelengths shorter than one meter. And the word "background" simply tells us that it can be measured everywhere.

What was predicted by the big bang theory was not just any microwave radiation at a few degrees above absolute zero. The prediction called for radiation that had the specific character of a *blackbody radiator*, pointing to radiation from a perfectly efficient heat source. Blackbody radiation is the radiation we would expect to find coming from an object that was perfectly black. Such objects are theoretical, because all objects reflect at least some of the light that falls on them. But by punching a hole in a box, we can create a hole that absorbs light almost perfectly. Both laboratory experiments and theoretical calculations tell us exactly what the spectrum of a blackbody radiator should look like.

This was the blackbody radiation that Alpher, Herman, and Gamow had predicted for the microwave background back in 1949, a type of spectrum that no object in the universe emits, but which we would expect for the entire universe. Such a blackbody radiation would provide evidence for the great amount of entropy (a measure of how much energy radiates from a system) that characterizes the universe itself.

Even the most entropic event ever observed, a supernova, contains only a tiny fraction of the kind of entropy we observe for the entire universe. Our second law of thermodynamics tells us that the universe is highly entropic, absorbing huge amounts of energy, leaving less and less energy available for work. Evidence of a blackbody background would provide specific evidence for an entire universe that began in a tremendous fireball of energy.

Thus the amount of radiation coming from a blackbody source should vary with wavelength in a particular way. If we measured the microwave background at a number of different frequencies and then plotted the radiation intensity against the wavelength, we should get what is known as a blackbody curve. Going from shorter wavelengths to longer ones, we would see that radiation intensity increases, then reaches a peak and falls off faster than it climbed, appearing like an uneven hump on a camel. Testing this microwave radiation to see whether it is truly cosmic in origin should be a matter of seeing whether it fits this very specific, spectral curve.

Early measurements fit this characteristic curve on the first side of its hump, but it wasn't until the mid-1970s that accurate measurements were made from a balloon and from a U-2 spy plane above the atmosphere to show that the curve also fell off in a way that fell almost precisely in line with the expected blackbody curve.

In 1987, however, cosmologists were thrown a curve into their curve. Physicists from Berkeley and Japan's Nagoya University collaborated on putting a spectrometer aboard a rocket. Too small to go into orbit, the rocket shot above the atmosphere long enough to collect just five minutes of data. This data, however, came back to show that the microwave spectrum diverged from the expected blackbody spectrum by *ten* percent at several wavelengths. Perhaps, speculated some theorists, the microwave radiation is not a relic from the big bang at all, but from some secondary blast that erupted all over the universe at a later time. Some wondered if the big bang theory was seriously flawed.

In 1990, NASA's $200 million COBE (Cosmic Background Explorer) satellite gave cosmologists the chance to make more accurate measurements. One of the three major instruments aboard (called FIRAS, the far infrared absolute spectrophotometer) had been developed especially to compare the microwave spectrum with the blackbody curve. John Mather headed the team that made this experiment capable of making measurements many times more sensitive than those of the Berkeley-Nagoya spectrometer. A few days after the 1989 launch, the protective instrument covers were blown away to uncover FIRAS.

A crowd of scientists gathered around the monitoring computers at the Goddard Space Flight Center just outside Washington, DC. All sensed that the shape of cosmology would be determined by the next few moments. Richard Isaacman's monitor flashed and then formed a perfect blackbody curve. Isaacman, who had helped calibrate FIRAS, began calling off the numbers as they came in from another computer. MIT's Ed Cheng plotted the points against the blackbody curve. The real test would be to see if the spectrum from space precisely fit the characteristic blackbody "signature" at all wavelengths. Cheng soon found that each number fell right along the blackbody curve. The Berkeley-Nagoya results had clearly been wrong. Isaacman's own belief in a creation event had been vindicated. Recalling the moment, he later said, "I felt like I was looking God in the face."[9]

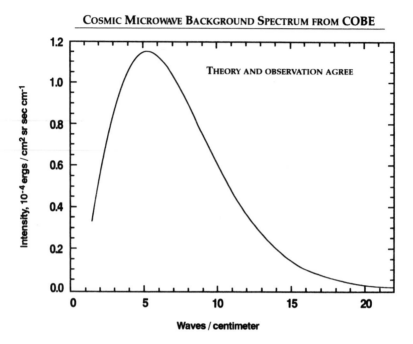

Cosmic Microwave Background Spectrum from COBE

The microwave background spectrum observed by the COBE satellite follows the blackbody curve perfectly. Deviations from the predicted blackbody curve are less than three-hundredths of one percent across the entire range of frequencies received by COBE. —*Courtesy of John Mather, Goddard, NASA*

The microwave radiation's perfect matching of the characteristic blackbody curve is now a scientific fact that is difficult to reconcile with anything other than a creation event involving the entire universe. University of Chicago's David Schramm reports that the FIRAS results "truly lent the greatest support to the Big Bang," since the blackbody spectrum "tells us that the universe was once so dense that it was a single, continuous body in thermal equilibrium, one that could be characterized by a single temperature."[10]

By 1990, John Mather's FIRAS team could announce that the cosmic background radiation deviated from a perfect blackbody radiator by less than one percent across the entire spectrum of observed frequencies. And in 1993, COBE measured the deviations at less than .03 percent.[11] This means that the big bang must account for at least 99.97 percent of the universe's radiant energy at wavelengths from ½ millimeter to 1 centimeter. At the end of 1994, John Mather, chief investigator of the FIRAS team, told me:

The big bang theory has passed the toughest test yet. Nobody has thought of another way for the universe to evolve that would produce such a perfect spectrum, although many scientists who like the steady state theory have tried very hard. All the alternatives require adding up little odd-shaped jigsaw puzzle pieces to make a perfect whole, and it doesn't seem very likely.[12]

Another reason that this microwave radiation is significant is that stars and galaxies simply do not emit microwave radiation at such high levels. As astronomers John Barrow and Joseph Silk explain: "There are no known sources bright enough at microwave wavelengths to account for this cosmic background. Furthermore, the background is so uniform in intensity from place to place that it would require more sources than there are galaxies in the entire universe to keep their random intensity fluctuations down at the observed level."[13]

Today the consensus of science is that the best explanation for the extremely even temperature in the blackbody radiation pouring in from all points of the universe is something that has all points of the universe as its source, something that produced far more energy than anything we have ever observed.

The Observed Elements

As mentioned earlier, the big bang theory predicts that the elements of the universe should exist in particular proportions: about three-quarters hydrogen, about one-quarter helium, and a tiny amount of all the heavier elements. Spectroscopic studies of the stars give us measurements in just these proportions.* In fact, without an event that is indescribably hotter than anything we know today, we cannot explain why helium exists in more than trace amounts, since it is impossible for helium to result in any great quantity from the burning of stars alone. Considering the fact that this discovery was first made in a study led by Fred Hoyle, who was looking for evidence to prove his steady state theory, these facts are not easy to dismiss.

The big bang theory predicts that the tremendous heat in the first few minutes of the universe's expansion should result in trace amounts of certain isotopes: deuterium (hydrogen with an extra neutron), light hydrogen, and lithium. Again, the observed relative abundance of these elements has not been adequately explained by any other theory. Heinz Pagels, Executive Director of the New York Academy of Sciences, explains the view of modern science: "Deuterium cannot be made in stars and survive . . . . The only viable explanation is that all the deuterium is primordial—it was made in the big bang."[15]

---

*To be precise, the big bang theory predicts that the light elements should have formed in the following proportions: 76 percent hydrogen, 24 percent helium-4, a few parts per 100,000 of deuterium, a few parts per 100,000 of helium-3, and just 1 part in 10 *billion* of lithium-7. Within the limits of what's possible to measure, observations fit these predictions very well. Even the trace amount of lithium (1 part in 10 billion) matches.[14]

Quasars and Distant Galaxies

(THINGS WERE DIFFERENT IN THE OLD DAYS)

To many of us quasars merely sound like a brand name. To astronomers, however, they are "bonfires on the shores of time," "open doors to the ends of the universe."

In 1960, Tom Matthews, a Caltech radio astronomer, noted something unusual about certain radio sources. They were tiny—mere points compared to the characteristic pair of lobes normally formed by radio galaxies. He sent his list of unusual radio sources to the man in charge of Mount Palomar's powerful 200-inch telescope, Allan Sandage. Having worked closely with Edwin Hubble in his later years, Sandage was considered his heir. Sandage succeeded in locating one of the tiny radio sources on a photographic plate. Galaxies, of course, had been routinely identified optically after first showing up as a radio source, but a single star had never been observed as a radio source before. Single sources lack the required power. Yet Sandage reasoned that it had to be a star, not a galaxy, because it varied greatly in brightness from one week to another, something galaxies could not do. For a galaxy to vary in brightness in this way would require billions of stars to brighten and dim in unison.

But the spectrum of this "radio star," as Sandage called it, turned out to look like no other star. It emitted abnormal amounts of energy in the blue and ultraviolet bands. No one could decipher the strange spectrum; it matched no known elements. Sandage's old Caltech teacher, Jesse Greenstein, momentarily considered the possibility that the spectrum represented a huge redshift, but this seemed like an impossible interpretation. Even the distant galaxies seldom exhibited such large redshifts, and no star had ever been redshifted to such a degree. Besides, such a large redshift would indicate that the star was at too great a distance to be picked up optically. No star was that bright.

In 1962, Caltech astronomer Maarten Schmidt solved Sandage's mystery for him when he recognized the clear pattern of hydrogen in the color readings. No one else had seen it because it was on the wrong end, far to the red side of the spectrum. Schmidt's calculations showed the object to be redshifted 16% (astronomers call this a redshift of .16), meaning that it was at least 1.5 billion light-years away and speeding away from us at a quarter of the speed of light.

Today some of these quasi-stellar objects (first dubbed "quasars" by a NASA engineer) are known to emit more energy than a thousand normal galaxies from a diameter only one trillionth the size of one average galaxy. They are by far the most powerful bodies we have observed in the universe, but their real mystique lies in the fact that they are the farthest objects we can see. Hence they also give us our earliest optical picture of the cosmos.

The questions naturally arise: Why are they all so far from us? And why do they become more numerous the farther we look into space (and hence the further we look back into time)? Evidently, the early universe contained an abundance of quasars.

Galaxies also look different the farther out we look. The most distant elliptical galaxies are apparently only about one-third to one-half the size of nearby elliptical galaxies; either that or it may be that only the stars in the core regions of these distant galaxies are "lit," making the galaxies appear smaller.[16] Also these older galaxies come in pairs about a third of the time, whereas nearby galaxies pair off just 7 percent of the time.[17]

Radio astronomers also find that the farthest galaxies (called radio galaxies) are more abundant, more densely distributed, than nearer galaxies.[18] These observations provide obvious evidence for a universe that has changed with time.*

The findings on quasar distances put one of the last nails in the coffin of the steady state theory. Most cosmologists believe that the observed distribution of quasars cannot be explained apart from the big bang theory.

Some astronomers believe that all galaxies somehow formed out of quasars; others view quasars as rare flukes in the universe. In either case, these distant beacons are giving astronomers the chance to observe other things that may settle the issue—like the huge clouds of hydrogen that lie directly between us and the most distant quasars. Quasars make the presence of these gas clouds known because the clouds absorb ultraviolet light from them, a characteristic of hydrogen gas. Initial observations show that there may be enough hydrogen gas in these most distant systems to account for all the material we now observe in galaxies. These clouds may have condensed to form the first generation of stars.[20]

Recent Developments—"Looking at God"

The same microwave background radiation that many consider the greatest evidence for the big bang has, until very recently, also been considered the greatest problem for the theory. When detectors were aimed at different parts of the sky, the microwave radiation appeared to be simply *too* smooth, too constant. It *should* be extremely even, just as our universe is extremely uniform (homogeneous, as the astronomers call it) on a large

---

*In September, 1994, researchers using Hawaii's 10-meter W.M. Keck telescope reported finding evidence that the more distant (earlier) cosmos had a microwave background temperature that perfectly matches the predictions of the big bang theory. The team used spectroscopic information to take the temperature of a distant gas cloud located in the path of light from a still-more distant quasar. Located at a distance that puts the cloud at 25 percent of the universe's current age, the cloud's temperature matches the big bang theory's predicted temperature for the microwave background radiation (7.58 kelvins) at that age.[19]

scale; but it should also be very slightly rippled (showing fluctuations of at least one part in 100,000) if it is to account for the slight inhomogeneity that had to be present at the beginning of the universe. This lack of uniformity at the beginning is necessary in all models of the big bang that account for the clumping of matter that could lead to galaxy formation. The old, standard big bang model has been shown to result in a universe with matter that is too evenly dispersed, a universe without galaxies.

The fact that we have galaxies today, and that they are organized in clusters rather than evenly distributed throughout the universe, means that this same local unevenness (inhomogeneity) must have been present at the beginning. However, this early inhomogeneity could not have been very great, or the universe would not have the overall uniform appearance that we observe in all directions today on a large scale. Matter is evenly distributed, but not *too* evenly distributed when you look more closely; otherwise the clumping of matter into galaxies would not be possible.

Somehow, the fluctuations that account for the ripples were just the right size, as Pagels notes, "big enough to eventually turn into galaxies but not so big that they would destroy the overall homogeneity of the universe."[21] Barrow and Silk agree: "Evidently, the balance between chaos and uniformity was rather delicate."[22] The precision of this balance, though inescapable, has been described as "unattractive" and "unaesthetic" to physicists, since it provides another argument for purposeful design.

And so many scientists began looking for other explanations in earnest, especially after evidence continued to pour in from COBE for an absolutely smooth microwave background. Though COBE carried the most sensitive instrument ever developed to detect ripples in the microwave background (called the DMR, the Differential Microwave Radiometer), preliminary data apparently showed the cosmic background to be as smooth as a Formica® table top. No bumps. No predicted ripples.

Some recent creationists seem to assume that atheistic scientists accept the big bang theory for philosophical reasons. That this is not the case should be clear from the fact that many gladly questioned the big bang theory when the COBE data began to give them reason to do so.

A number of the articles and books I read between 1990 and 1992 claimed that the standard cosmological model failed miserably in explaining how matter could possibly coalesce into galaxies from a perfectly uniform explosion. Worse, the new star maps that came out at this time, based on the observations of Margaret Geller and John Huchra, clearly showed that galaxies are clumped together in massive clusters and superclusters. If the microwave background radiation was truly the remnant of a big bang, it simply had to contain the seeds of this colossal clumping. Recent creationists delighted in quoting from the flood of articles written by big bang doubters during this period.

But on April 24, 1992, astrophysicist George Smoot finally announced that the COBE satellite had measured the expected "ripples" in the microwave background radiation. Actually, Smoot and his team had discovered the clear signal many months before. One team member in particular, Ned Wright, wanted to publish the findings and had a running argument with George Smoot about it. But Smoot kept his team from leaking any information on the growing picture that formed from each day's data until COBE's Science Working Group had ruled out every other conceivable cause.

Though the fluctuations in the background were smaller than some had anticipated, computer analyses proved that they were definitely there (at about 1 part in 90,000; or to be precise, 11 ± 3 parts per million),* and they were clearly distinguishable from noise. Newspapers all over the world carried Smoot's quote: "If you're religious, it's like looking at God."[23]

Stephen Hawking, whose own work had recently sought alternatives to a universe with a beginning, couldn't have been more impressed with the findings. He called them "the most important discovery of the century, if not of all time."[24]

The COBE's DMR results were soon checked against those obtained by the high altitude balloon tests performed by University of Chicago's Stephen Meyer. His December 1992 findings agreed closely with the level of fluctuations found by COBE. Before NASA shut it down at the end of 1993, COBE performed a final year's worth of data-gathering, refining and confirming the previous years' measurements.

The findings added evidence, not only to the framework of standard big bang cosmology, but to the impression that our universe began in an incredibly well-organized manner, a manner that almost seemed orchestrated in a way that would distribute matter for our benefit, allowing for the formation of stars and galaxies. George Smoot later wrote: "And when we see that a noticeable tilt in the quantum fluctuation spectrum of primordial ripples might have produced, instead, a vast swarm of black holes, or a cosmos of lumbering giants, then we again realize how easily things might have been very different."[25]

The ripples now observed in the cosmic background radiation give us a picture of what our universe looked like at the earliest moment we can directly observe, at the time when our universe first became transparent.

---

*Actually, some models of galaxy formation would have allowed for much smaller fluctuations than COBE was capable of detecting. Thus the big bang theory was never in as much trouble as some dissenters tried to make out.

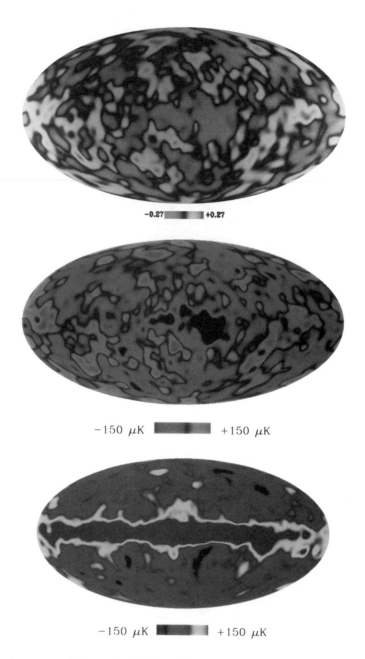

−0.27 ▮▮ +0.27

−150 μK ▮▮▮ +150 μK

−150 μK ▮▮▮ +150 μK

Three sky maps made from the COBE satellite's DMR instrument show faint ripples (deviations) in the temperature of the cosmic background radiation. Though the fluctuations are barely perceptible above the instrument noise, computer analysis clearly indicates their presence. Differences in color represent deviations in cosmic radiation temperature of one hundredth of one percent above or below the average temperature of 2.73 degrees above absolute zero. Top: Sky map from first year of DMR data. Center: Same map, refined from two years of data. Bottom: Same two-year sky map, but here the prominent emission from the plane of our Milky Way Galaxy has not been reduced. —*Courtesy of John Mather, Goddard, NASA*

George Smoot (left) at work with lab assistant, Rich Bell, at Lawrence Berkeley Laboratory. —*Courtesy of Lawrence Berkeley Laboratory, University of California*

"Looking at God"

Interview with George Smoot[26]

Heeren: When the COBE satellite first measured the fluctuations in the background radiation, you made the much publicized statement: "If you're religious, it's like looking at God." Could you explain something about the basic implications of the big bang theory and what you meant by that statement?

Smoot: Well, I meant there were two aspects to it. These were the oldest and largest structures ever seen. Not only did we find what are the seeds of the modern day structure—and that is the galaxies and clusters of galaxies and clusters of clusters of galaxies—but we also found evidence of the birth of the universe, I believe, because I think that if you look at these fluctuations and ask, "How could they have gotten in there?," some of them are so large—that is, they stretch across billions of light-years back at a very early time —that means they hadn't changed—if you move matter and energy around at the speed of light, you can only cross a teeny fraction of them. And so these are primordial—they're in from the moment of creation.

And so it's really like looking back at creation and seeing the creation of space and time and the universe and everything in it, but also the imperfections of the creation, sort of the fingerprints from the Maker, if you understand what I mean, or the machining marks from the machine that tooled the universe, and those things very neatly turn out to be the things that caused the universe to be very interesting to us: namely, creating galaxies and stars and so on. So, to me, the implications were really quite profound, the idea that not only do we understand where things came from, but those things were actually like the machining marks, the manufacturing marks, from the creation of the universe.

Heeren: And these had to be very precise—they had to be within very tight boundaries in order to produce anything that made any— to make the right kind of preparations for life.

Smoot: Right. In order to make a universe as big and wonderful as it is, lasting as long as it is—we're talking fifteen billion years and we're talking huge distances here—in order for it to be that big, you have to make it very perfectly. Otherwise, imperfections would mount up and the universe would either collapse on itself or fly apart, and so it's actually quite a precise job. And I don't know if you've had discussions with people about how critical it is that the density of the universe come out so close to the density that decides whether it's going to keep expanding forever or collapse back, but we know it's within one percent.

The Quest for God's Fingerprints

NOTE FROM CARL: ONCE AGAIN, THOSE WHO ARE LESS SCIENTIFICALLY INCLINED MAY WISH TO SPARE THEMSELVES FROM THIS SECTION'S SOMEWHAT TECHNICAL EXPLANATION OF THE FACTS BEHIND THE QUEST.

George Smoot had earlier pioneered the making of incredibly fine measurements of the microwave background radiation by sending his differential microwave radiometer (DMR) up in the U.S.'s old high altitude spy plane, the U-2. His observations in the late 1970s were among the first to show that galaxies are not distributed fairly evenly throughout the universe, as previously thought.

Six months of consistent results from his U-2 flights showed that our entire galaxy is being pulled through space at more than a million miles an hour by a great unseen force. Today we know that galaxies are grouped in clusters and in "clusters of clusters" (superclusters). The velocity of our Milky Way (along with a dozen neighboring galaxies known as the Local Group) relative to the microwave background radiation is due to the gravitational tug of a distant, massive structure known simply as "the Great Attractor" in the direction of the Hydra-Centaurus supercluster. The fact that such enormous structures exist in the universe led Smoot to realize that the cosmic seeds that produced them must have been present at the birth of the universe. And so began his search for the cosmic seeds.

In his classic text, *Physical Cosmology*, Jim Peebles had predicted that we should be able to measure the motion of our galaxy relative to the microwave background radiation, using this cosmic radiation as a universal frame of reference. Such an absolute frame of reference sounds something like the old concept of an "ether," a theoretical universal medium earlier scientists had sought to account for the passage of light through space.

The famous Michelson-Morley experiment had disproved the existence of ether in 1887, and Einstein had showed in 1905 that all motion is relative; there *is* no universal frame of reference. However, Peebles reasoned that if the earth is actually moving relative to the cosmic background radiation, then it might be possible to detect a Doppler effect, using the cosmic radiation as if it were the fabled ether. The temperature of the cosmic background radiation should be slightly warmer in the direction in which the earth is moving, cooler in the direction of recession (like rain hitting the front windshield of a moving car harder than the back windshield).

George Smoot's precision instruments first detected this effect in the microwave background radiation, called dipole anisotropy (two poles: one warm and one cool), while aboard the U-2. He developed an even more sensitive version of his DMR detector to use aboard NASA's COBE satellite in order to detect the smaller ripples in the cosmic background radiation.

After his first results with his instrument aboard the U-2, Smoot wrote: "The big bang, the most cataclysmic event we can imagine, on closer inspection appears finely orchestrated."[27]

These ripples in the microwave background radiation (also called "cosmic seeds" or "fingerprints of God") were predicted by all the models currently proposed for the big bang, versions of the theory that are collectively called the inflationary model.

The Inflationary Big Bang

The inflationary model of the big bang theory, first put forth by Alan Guth (whose name rhymes with "booth") of MIT in 1979, is the one scenario that is apparently consistent with all observations—at least for now. According to its proponents, the universe can be expected to undergo a brief phase in which the initial expansion rate speeds up exponentially,* in some models lasting only from 10^{-36} to 10^{-34} second after the universe begins.

According to Guth's inflation model, this acceleration can be attributed to the transition that takes place when falling temperatures make it possible for the first particles to permanently form. It has been compared to the transition ice undergoes when it melts; the ice molecules release once-latent energy as they take their place in a new, less rigid state. In the same way, the new universe may have experienced an energy-releasing transition, briefly accelerating its expansion, and then, having spent the transition's energy, the universe returned to the much less rapidly-paced, decelerated expansion predicted by Friedmann, which we now observe.

The size of the universe at the end of this transition is calculated to have been about the size of a baseball. Less than a trillionth of a second later, when its radius had grown to about three feet and its temperature had "cooled" to about 100 billion degrees, the constituents of atoms had formed: quarks and electrons. Atoms themselves formed when the universe was about a millionth of a second old. Quarks clumped together to form protons and neutrons, protons and neutrons clumped together to form nuclei, and soon helium, the first element, was flying through the intense radiation that we still detect today, at greatly reduced levels, everywhere in the universe.

Without inflation, we have no explanation for why this radiation is so uniform in temperature from one region of the universe to the other, since heat would have had to travel faster than light (an impossibility) in order to flow across these regions in the small amount of time available. Yet, satellite observations show the opposite sides of our visible universe to have the same cosmic background temperature (to one part in almost 100,000). Clearly, these widely separated portions of the universe were once in contact. Robert Jastrow explained the problem to me this way:

> How do you account for the fact that parts of the universe—at opposite sides of the visible universe—which means they're 30 billion light-years apart—could not have been in contact? It would take too long for light to

---

*Dr. Guth explains: "'Exponential expansion' just means that there's a fixed time period over which the size of the universe would double, and in the next period of the same length the size of universe would double again, and then it would keep doubling, each successive time interval."[28]

Inflationary Big Bang

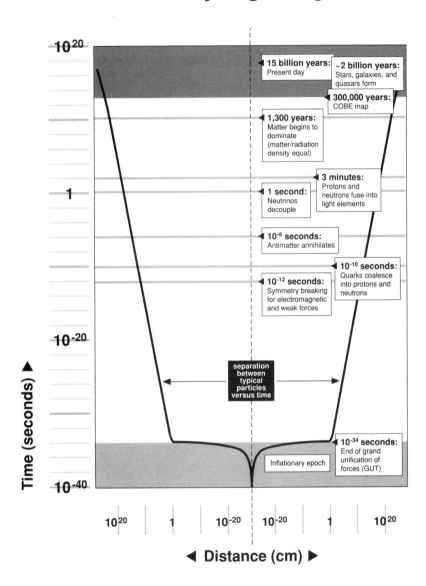

—*Illustration © 1995 Christopher Slye*

cross from one to the other; yet there are almost exactly the same levels of cosmic background radiation on both sides of the universe. I'd suggest that they exchanged energy at some time in the past—but they couldn't have. But the inflationary universe, with its rapid initial growth, solves that problem, because they could have been in contact at the very beginning.[29]

In other words, the big bang theory takes us back to a time when, after the first moment of creation, the entire universe consisted of a region a trillionth the size of a proton. In this first instant after creation, the universe may have been in a homogeneous state. All parts were in contact with and perhaps even equivalent to one another. Inflation theory tells us that, in a tiny moment after this first instant of creation, the universe rapidly expanded by a factor of a hundred times the amount by which it has expanded in the subsequent 15 billion years. During this sudden surge, the homogeneity that characterized the universe when it was of subatomic size became the characteristic of the entire universe in post-inflation times.

Alternatively, any unevenness in temperature that might have existed prior to inflation could have been smoothed out by the rapid force of expansion. Either way, inflation gives us a solution to explain how these widely separated parts of the universe were initially in contact.

John Mather and George Smoot have told me that inflation matches well with their analysis of certain COBE satellite data. Inflation predicts a particular distribution of the microwave background ripples, known as the scale-invariant power law fluctuation spectrum. What this mouthful means is that if inflation occurred, the COBE data should show equal amounts of small, medium, and large-size ripples, no matter which slice of the sky we observe (as long as the slices are equal in size). Other, earlier models also predicted this type of distribution of the ripples, however; so this prediction is not exclusive to inflation.*

The newer inflationary models produce all these effects without the phase transition earlier mentioned. Russian-born physicist Andrei Linde is now proposing

Alan Guth. —*Courtesy of Donna Coveney/MIT*

---

*The slices of the sky measured by COBE were far too large to tell us anything about the distribution of early ripples on small angular scales. The COBE data cannot pinpoint the formation of anything even as small as the largest structures we know about (like the massive superclusters of galaxies known as the "Great Wall" or the "Great Attractor"). Newer groups of researchers taking measurements from the South Pole, from high altitude balloons, and from radio telescopes have begun to report on the fluctuation levels at angular scales below 2 degrees, something COBE could not do.

inflation based on certain theories of elementary particles. According to Linde, the importance of inflation theory has itself inflated so that "inflation is not a part of the big bang theory, as we thought 15 years ago. On the contrary, the big bang is a part of the inflationary model."[30] Even Linde, however, admits that "the possibility exists that some new observational data may contradict inflationary cosmology."[31] Inflationary cosmology is still a very young science.

Interview with Alan Guth[32]

NOTE FROM CARL: THIS SECTION, ONCE AGAIN, IS MAINLY FOR THOSE WHO ARE INTERESTED IN THE TECHNICAL EXPLANATIONS BEHIND THE DISCOVERIES.

Heeren: How do the newest inflation models differ from your first one?

Guth: My original model actually was not a successful one; it failed in the end. I assumed that when the phase transition that we've been discussing would eventually take place, that it would take place in what really is the most common way for such phase transitions to take place, in a manner similar to the way water boils.

The assumption that I was making was that the phase transition occurred by bubbles of the new phase forming and then starting to grow. That turned out not to work because of the fact that when the bubbles form, they form randomly, and one of the goals of inflation was to explain why the universe looked so uniform. And this exponential expansion process actually does a beautiful job of producing an incredibly smooth and uniform universe, which is what one wants to explain the real universe. But nonetheless, when the bubbles start to form randomly, it turned out they created gross nonuniformities in space and completely destroyed all the beautiful uniformity that was created by the exponential expansion.

So the model as I produced it was an attractive model, but a model that clearly wasn't quite right. The first successful version of inflation was invented about two years later independently by a Russian physicist, Andrei Linde, and two Americans, Albrecht and Steinhardt.

Those readers who are already familiar with new developments in cosmology know that theorists have found inflation useful in solving a great many problems. But what intrigued me most was the suggestion by some that inflation even solves the problem of explaining how the universe itself could have come out of nothing.* Perhaps such a claim should not worry the Bible believer; after all, if God used natural processes to form His universe, He may have used a natural process to create it out of nothing in the first place (as at least one Bible-believing physicist I have talked to believes[33]). However, when I found that some were taking the statements of very prominent physicists to mean that inflation theory does in fact explain how the universe came out of nothing, I put on my skeptic's hat. I decided to go right to the scientist who was closest to inflationary theory, the man who developed it, Alan Guth, with this question.

Heeren: I get the impression that many physicists interpret your equations for inflation to give a full explanation for how the universe and space itself came out of absolutely nothing. Do *you* feel that inflation explains how the universe came out of absolutely nothing?

Guth: First of all, I will say that at the purely technical level, inflation itself does not explain how the universe arose from nothing. . . . Inflation itself takes a very small universe and produces from it a very big universe. But inflation by itself does not explain where that very small universe came from.

Dr. Guth used the term "inflation *itself*" because he acknowledged that there are many ongoing efforts to find a natural explanation for how the universe could have originated from absolutely nothing. Inflation is an important tool in these efforts; however, the inflation theory itself cannot make the claim of explaining how something can come from nothing. Dr. Guth also gave several examples of fine-tuning problems that inflation cannot solve (that is, inflation requires several critical parameters that have no natural explanation; see Bonus Section #2). The more I speak with scientists who are convinced that there are naturalistic explanations for *everything,* the more it seems to me that these scientists are actually finding considerable evidence of fine-tuning (intelligent design) within their own fields. Everyone seems to think that the evidence for a nontheistic view of the universe lies in someone else's specialty.

Even Andrei Linde, the most flamboyant inflationary cosmologist (who claims he may be able to create an entire universe in a laboratory if he can just figure out how to trigger inflation),[34] admits that inflation can't explain how something could come out of nothing. He writes: "Explaining this initial singularity—where and when it all began—still remains the most intractable problem of modern cosmology."[35]

---

*Here are two examples of how physicists make statements that can be taken to suggest this. In our May 13, 1994 interview, Robert Gange told me: "If you look at the Alan Guth equations out of MIT that introduced inflation into the big bang, you will find that those equations can be understood to describe a creation out of nothing." And in our May 6, 1994 interview, George Smoot told me that "inflation is a model that explains how we create space and time and everything that's in space and time."

New discoveries, better sky maps, and finer measurements of the background ripples should continue to revise big bang cosmology in the coming years, but the consensus today is that some model falling within the standard big bang framework will best explain all observations. And the most widely accepted big bang models now describe, not a random explosion, which could never have produced the galaxies we observe, but a precisely controlled beginning for the universe (as we will see in more detail in Chapter 9).

Big Bang Expansion: An Orderly Progression

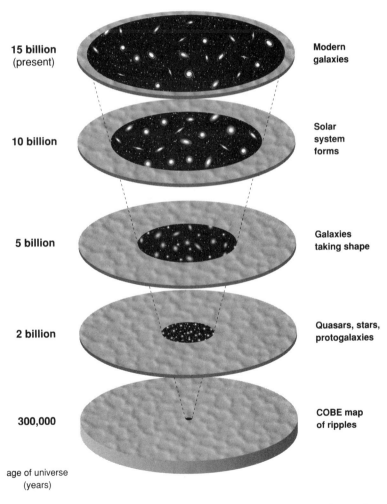

15 billion (present) — Modern galaxies

10 billion — Solar system forms

5 billion — Galaxies taking shape

2 billion — Quasars, stars, protogalaxies

300,000 — COBE map of ripples

age of universe (years)

The big bang could not have been a random explosion. Rather, all observations point to an incredibly finely-tuned, carefully orchestrated chain of events that could lead to the solar systems necessary for intelligent life. —*Illustration © 1995 Christopher Slye*

Such a finely controlled creation event (and for that matter, *any* kind of a creation event) defies both atheism and all the religions of pantheism, since both require an eternal universe. Both require that the universe somehow be its own cause, a logical impossibility in any case, but more clearly so if the universe had a beginning. By 1927, the world of science had already been greatly shaken by the news of redshifting galaxies and Lemaître's "expanding universe" solutions to Einstein's equations. The evidence for a universal beginning led Sir Arthur Eddington to state, "Religion first became possible for a reasonable man of science in the year 1927."[36]

In contrast to the naturalistic tendencies of 19th-century science, the evidence has moved scientists to take a distinct turn from the non-creationist view held before the 1920s. Today, George Smoot writes: "The question of 'the beginning' is as inescapable for cosmologists as it is for theologians."[37] And one of our most prominent astronomers, Robert Jastrow, notes that "the essential element in the astronomical and biblical accounts of Genesis is the same; the chain of events leading to man commenced suddenly and sharply, at a definite moment in time, in a flash of light and energy."[38]

Shocking Summary Statements
and
Stimulating Conversation Starters

⚡ Some creationists believe in the big bang theory and some don't; but everyone who believes in the big bang theory must be a creationist.

⚡ The standard cosmological model of modern science describes a creation event that defies atheism and pantheism, but harmonizes with the Bible.

⚡ Christians shouldn't feel threatened by modern cosmology. The big bang theory, after all, explains nothing about ultimate origins. It only raises questions that science can't answer.

⚡ Edwin Hubble found that the farther galaxies (which show us the more distant past) are retreating from us faster than the nearer galaxies, just as one would expect if the universal expansion is slowing down from an initial surge. Thus observation tells us that the universe had a beginning. Famed astronomer Robert Jastrow says: "The Hubble Law is one of the great discoveries in science: it is one of the main supports of the scientific story of Genesis."

When Penzias and Wilson stumbled upon the microwave background radiation in 1965, they fulfilled the prediction of early big bang theorists that such a low level, universal radiation should exist. But more important, the spectrum of this cosmic radiation has since been measured to perfectly match the characteristic blackbody curve. This precise matching is difficult to reconcile with anything other than a creation event involving the entire universe.

Without an extremely hot, dense creation event, the large amount of helium evident in the universe today cannot be explained.

The fact that galaxies are distributed more densely—and quasars become abundant—the farther we look into space indicates that the universe has changed with time. These observations argue against any static or steady state theory and for a creation event.

After the COBE satellite team discovered the predicted ripples in the cosmic microwave radiation, George Smoot called these fluctuations "the fingerprints from the Maker." Smoot draws attention not only to the fact that his team had provided more evidence for the creation event, but for a "finely orchestrated" creation event. Stephen Hawking was so impressed with this finding that he called it "the most important discovery of the century, if not of all time."

There has never been a time in history when humankind has had such an unmistakable message from space to tell us that there is a super-intelligent, transcendent Creator.

Notes for Chapter 7

1. Robert Jastrow, *God and the Astronomers*, second edition (New York & London: W.W. Norton & Company, 1992), p. 93.

2. George Smoot and Keay Davidson, *Wrinkles in Time* (New York: William Morrow and Company, 1993), p. 30.

3. Jastrow, p. 53.

4. Robert Jastrow, interview with author, June 7, 1994. Dr. Jastrow made the specific point, "I'm an agnostic," twice during the interview.

5. Fred Hoyle, *The Intelligent Universe* (New York: Holt, Rinehart and Winston, 1983), p. 237.

6. John D. Barrow and Joseph Silk, *The Left Hand of Creation—The Origin and Evolution of the Expanding Universe* (New York: Basic Books, 1983), p. 45.

7. Arno Penzias, interview with author, May 4, 1994.

8. Robert Wilson, interview with author, May 18, 1994.

9. Richard Isaacman, quoted by John Boslough, *Masters of Time—Cosmology at the End of Innocence* (New York: Addison-Wesley Publishing Company, 1992), p. 45.

10. David Schramm, "Dark Matter and the Origin of Cosmic Structure," *Sky & Telescope* (October, 1994), p. 29.

11. Ron Cowen, "COBE: A Match Made in Heaven," *Science News*, vol. 143, no. 3 (January 16, 1993), p. 43.

12. John Mather, in letter to the author, December 14, 1994.

13. Barrow and Silk, p. 17.

14. Schramm, p. 29.

15. Heinz R. Pagels, *Perfect Symmetry* (New York: Simon and Schuster, 1985), p. 154.

16. Ron Cowen, "Tracking the Evolution of Galaxies," *Science News*, vol. 145, no. 5 (January 29, 1994), p.77.

17. Ibid.

18. Smoot and Davidson, p. 77.

19. Ron Cowen, "Taking the Temperature of the Far Cosmos," *Science News*, vol. 146, no. 11 (September 10, 1994), p. 166.

20. Ron Cowen, "The Debut of Galaxies," *Astronomy* (December 1994), pp. 50, 53.

21. Pagels, p. 132.

22. Barrow and Silk, p. 108.

23. Milton Rothman, "What Went Before?" *Free Inquiry*, vol. 13, no. 1 (Winter, 1992/93), p. 12.

24. Stephen Hawking, quoted by Smoot and Davidson, p. 283.

25. Smoot and Davidson, p. 190.

26. George Smoot, interview with the author, May 6, 1994.

27. Smoot and Davidson, p. 135.

28. Alan Guth, interview with the author, June 8, 1994.

29. Robert Jastrow, interview with the author, June 7, 1994.

30. Andrei Linde, "The Self-Reproducing Inflationary Universe," *Scientific American* (November 1994), p. 55.

31. Ibid., p. 53.

32. Alan Guth, interview with the author, June 8, 1994.

33. In my May 13, 1994 interview with Robert Gange, a Bible-believing physicist, he clearly stated this view.

34. Linde, p. 53.

35. Ibid. p. 48.

36. Arthur Eddington, quoted by David Foster, *The Philosophical Scientists* (New York: Dorset Press, 1991), p. 1.

37. Smoot and Davidson, p. 189.

38. Robert Jastrow, *God and the Astronomers,* second edition (New York & London: W.W. Norton & Company, 1992), p. 14.

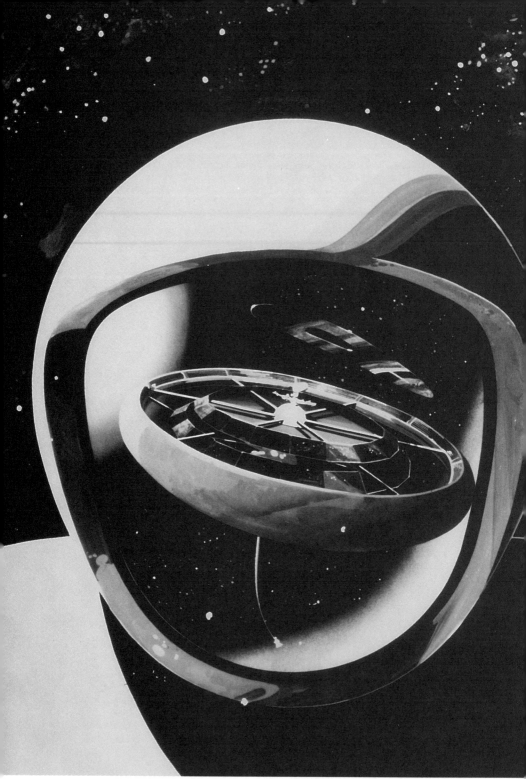

As seen through a future astronaut's helmet: NASA has been toying with plans for a space colony that will accommodate a manufacturing workforce of 10,000 people. Alternatively, the author thinks it might be a nice place to put people who wish to devote themselves to a lifetime of entanglement in the recent creation/old creation controversy. —*NASA*

CHAPTER 8

The Bible and the Big Bang

He has made everything beautiful in its time.

Ecclesiastes 3:11a

Carl: One thing I noticed in that last chapter: you're a rotten name-dropper. If you're going to drop names, try to at least pick some names that people have heard of. Who's heard of all these physicists? Try some rock stars, maybe a few football players.

Fred: What do rock stars and football players know about the origin of the universe?

Carl: What difference does it make? They'll have opinions. People would like to know what they think. Maybe you could get Arnold, uh . . .you know, *Arnold*. What's Arnold's last name?

Fred: Arnold who?

Carl: Bruce Willis!

Fred: Did you mean Arnold Schwarzenegger?

Carl: Yeah, get an action hero to comment on the cosmos. You'll sell out your first edition the first day. How's this for a concept? Let Arnold and Bruce and Sly Stallone slug out the whole controversy about the age of the universe. Arnold can be a conservative, recent creationist and he can take on Bruce and Sly, the old universe creationists.

Fred: Maybe I'm just naive, but I don't see why the disagreement between recent creationists and old earth creationists is worthy of any great controversy.

Carl: Controversy sells! We'll call it the battle of the ages! The battle *about* the ages!

Fred: Some of the great Bible teachers who I respect most happen to be recent creationists, though I disagree with their position on creation's timing. I just don't think they're aware of all the facts. Science isn't their field.

Carl: It's not Arnold's field either. But I still think more people will listen to him talk about it than some old Nobel prize-winning physicist they've never heard of.

Fred: Forget the controversy angle. The fact is, the age of our universe is not spelled out in the Bible. Most of what the Bible talks about is what's been happening for the last 4,000 years, which young and old universe proponents can agree upon. I can still learn a lot from Bible teachers who disagree with me on this point. The *fact* of creation is critical, but God's timing is not something to break fellowship over.

A Skeptic's Questions:

How am I supposed to take you Bible believers seriously when so many of you say that the universe was created within the last 10,000 years, in obvious contradiction to the facts of science?

A Bible Believer's Response:

The fact that so many Christians believe that may be more of a cultural phenomenon than any sign of what the Bible actually teaches. I won't deny that there's a conflict between science and traditional beliefs among many Christians, but as I can show, there's no conflict with the Bible itself.

Conflict with Recent Creation Tradition

For Christians who wish to pay attention to science, science has brought them good news and bad news. The bad news, for some, is that the universe appears to have been created billions, not thousands, of years ago. The *good* news is that it was *created*.

For the healthy skeptic, the consensus of science is open to question and independent examination, since it is based on the tradition of human beings, who are never infallible. Some Christian readers may feel relieved to know that I will treat the subject of evolution (in Volume 2) with far greater skepticism than this treatment of the big bang theory, and with good reasons from the evidence.

But if the consensus of scientists is open to question because of their susceptibility to human error, might not the tradition of many Christians also be open to error—particularly when Bible-believing Christians disagree with one another—and most particularly when their traditional opinions are not based on the clear teachings of Scripture, but on a personal interpretation?

We have mentioned scientists who, because of their theological biases, did not accept the idea of a beginning for the universe; but to the credit of most, when the evidence became overwhelming, their desire for truth won them over to what was for them a new way of thinking. This sums up the complete turnaround of science in this century from a belief in an eternal universe to a universe with a creation event. My own theological biases long included the idea that God created, not just transcendently, but *recently*. For many of us, our unwillingness to even consider a theory that suggests otherwise is based, not on scientific evidence, and not on our own diligent study of this matter in Scripture, but on our faith in an interpretation of certain Christians.

The Source of Recent Creation Tradition

Recent creation tradition stems especially from James Ussher (1581-1656), Irish Archbishop of Armagh. Genealogies of Genesis and other portions of the Bible led him to the conclusion that there were 4,036 years from the creation of the universe until the birth of Christ. Not satisfied with this, scholar John Lightfoot (1602-1675) took it upon himself to refine the calculations, and he decided upon a creation date of October 18, 4004 B.C. Adam was created on October 23 at 9:00 a.m.

Creation critic E.T. Brewster wryly comments: "Closer than this, as a cautious scholar, the Vice-Chancellor of Cambridge University did not venture to commit himself."[1]

These chronologies were based on the mistaken notion that the Bible's genealogies always included every generation. Neither Ussher nor Lightfoot recognized the fact that, in the Old Testament, the Hebrew words for father (*'ab*) and son (*ben*) can also mean forefather and descendent. Bible scholars (including even the staunchest recent creationists) now recognize that "telescoping" was a common practice, serving as an aid to memory and as a means of emphasizing the more important ancestors. The practice becomes especially obvious to any scholar who examines passages such as 1 Chronicles 26:24, Ezra 7:2, or Matthew 1:8.

But this date for creation quickly became a part of Protestant tradition when it was included as a margin note, and sometimes even as a heading, in the early printings of the authorized King James Version of the Bible.

Explaining a Young Universe

These dates, or at least the general time periods, became ingrained into much Christian thinking, but they posed no real problem until conflicts arose with geological findings in the early 19th century. Today, the young earth position is in direct conflict with practically all the sciences. Recent creation advocates must explain how we can see galaxies that are billions of light years away (meaning their light took billions of years to get here). Fossil evidence and radioactive dating of the rocks themselves have rendered any recent creation theory increasingly unreasonable.

Philip Gosse, a Christian geologist who found himself confronted with undeniable evidence for an ancient earth in the 1850s, resolved his dilemma by proposing that God had created everything with an *appearance* of great age. The actual age of the earth might be eight or ten thousand years, but the "ideal" age (the age which God simulated at creation) could be millions of years. Ancient fossils, according to this view, never existed as plants or animals but had been *created as fossils*. In His exhaustive job of "antiquating" His universe, God had even simulated the half-digested food found in many animal fossils.

Of course, an omnipotent God has the ability to do such a thing. But the question arises, Would the God who encourages us to delight in and study His great works create false histories (Psalm 111:2)? Are His creative wonders deceptive, or do we appreciate Him more the more we learn of His lengthy preparations for humankind?

The "creation-with-apparent-age" hypothesis is still held by most recent creation advocates today. According to it, we see starlight, not because the stars were really there billions of years ago, but because God created the starlight *in situ*, as if it had been traveling for billions of years. Such a situation would mean that events that we see in distant galaxies, such as supernovas, never actually happened, but that God is sending us false stellar reports by light. Alternatively, some propose that the speed of light has actually changed. Perhaps light used to travel with much greater velocity, quickly bringing us the light from billions of light-years away, and then God slowed it down to its present velocity.

Such reasoning, of course, is based on questionable presuppositions, not on evidence. Even science writer Paul Steidl, a recent creationist who prefers this view, admitted, "To be honest, there is little or no evidence for a change in the speed of light."[2] There is, however, much contrary evidence. More than fifty experiments to measure the speed of light, beginning over three hundred years ago, indicate that this constant has not changed during that time (at least, not as far as we can tell, considering the imprecision of data used in the past).[3] More importantly, astronomers who measure the spectral lines of hydrogen report no change in the light waves arriving from even the most distant galaxies. The spectral line of hydrogen at 21 centimeters should vary with the speed of light, but this line remains constant, whether the measurement is made on nearby galaxies or those billions of light-years distant.[4]

Another early idea of the recent creationist advocates, called the "tired light" hypothesis, calls into question the whole concept of determination of distances by redshifting. According to this idea, redshifting may be due to the light's loss of energy rather than the Doppler effect. But careful measurements show that identical redshifts are observed even at greatly disparate frequencies; any loss of energy from distant starlight would show increasing amounts of blurring across spectral lines.[5] In spite of the fact that the "tired light" hypothesis was tested and proved groundless years ago, recent creation advocates continue to promote this hypothesis today.[6]

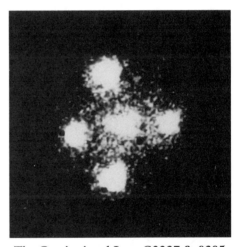

The Gravitational Lens G2237 & 0305. Sometimes referred to as "the Einstein Cross," this Hubble Space Telescope image shows a very distant quasar that has been multiple-imaged (the four outer objects) by a relatively nearby galaxy (the central object) acting as a gravitational lens. Gravitational lensing occurs when light from a distant source passes through or close to a massive foreground object. The number and placement of images depends upon the detailed alignment of the foreground and background objects with the line of sight to earth. Snapped by the European Space Agency's Faint Object Camera aboard HST, this is the most detailed image ever captured of a gravitational lens. —*NASA*

Those who question the whole idea of redshifting and an expanding universe should understand that gravitational lensing independently confirms that objects with higher redshifts truly are more distant. Massive objects like galaxies are known to act as crude lenses, bending the light from objects that lie behind them so that we can see them.

Gravitational lenses thus distort the images of these objects, magnifying them or producing multiple images of the same object. In every case where a gravitational lens has been located, the object behind the lens always has a higher redshift, confirming that the redshift does indeed indicate greater distance.

Other factors do contribute to the degree of an object's redshift besides the Doppler effect (like gravitational effects). But even if new evidence suddenly came along to show that the redshifting now attributed to recession velocity should be attributed in great degree to something else, we still know that the farthest observable galaxies are at least some billions of light-years distant, and so we know that the universe is at least a few billion years old.

How can we be so sure? Given the distances we have determined to our nearest stars by fool-proof parallax methods, and given a minimum amount of spacing between stars, we know that our own galaxy's size must be close to the 100,000 light-years that we have determined by other methods. Even if we suppose that there is very little space between galaxies, we have to account for the space required by the millions of observable galaxies (to be extremely conservative, over a million have been observed and plotted on maps—astronomers say there are actually billions). Even if the typical galaxy is a tenth the size of ours, and even if we assume there is no space between galaxies at all, and even if we assume there are only 100,000 galaxies, we find our observable universe must be at least a billion light-years in size (10,000 light-years per galaxy x 100,000 galaxies). To be able to receive light coming from such distances obviously requires that the universe be at least a billion years old.

Even recent creationist Paul M. Steidl writes:

> There is no doubt that there really are objects greater than 10,000 light years away. Ten thousand light years still leaves us well within the Milky Way Galaxy, and other galaxies are thousands of times farther away than this. No arguments or revisions of the distance scale can change this basic fact.[7]

To accept the idea that God created starlight "in transit" and rocks with radioactive elements used up in order to give them the "appearance" of billions of years of age is to accept the idea that God is deceptive. At best the appearance-of-age hypothesis presumes that God has some hidden agenda, some reason to misrepresent the universe as much older than it is. Even Einstein's notion of God was above this; he refused to believe that there was anything deceptive about the way He created the universe: "Subtle is the Lord," said Einstein, "but malicious He is not."[8]

To accept the idea that starlight has slowed down over the years or that its light has become "tired" over distance is to ignore laws of physics and observational tests. To seriously cling to any of these ideas is to put preconceptions (which are based more on tradition than on Scripture) above evidence.

Christian philosopher of science Bernard Ramm says: "Such a scheme as Gosse propounds, clever as it is, is a tacit admission of the correctness of geology. Better sense will state that the ideal time is the real time."[9]

The Other Christian Tradition

For those who feel the need for support from Christian tradition, I should point out that Augustine said that the creation "days" were not sun-divided days, but rather, God-divided days.[10] Others who believed that the creation days were not necessarily solar days include Irenaeus, second century apologist and martyr; Origen, third century apologist; Basil, 4th century bishop of Caesarea; and Thomas Aquinas, 13th century theologian.[11] This tradition even goes back to ancient Jewish thinking, for the position that the days of Genesis were not to be taken literally was also held by first-century Jewish writers, Philo and Josephus.[12] Obviously, these early opinions cannot be said to have been formed in order to comply with the discoveries of modern science.

More recently, the ancient universe view was held by conservative theologian C.I. Scofield, whose Reference Bible helped publicize the famous "gap" theory (which proposed an age-old earth before Adam). Scofield wrote: "The first creative act refers to the dateless past, and gives scope for all the geologic ages. . . . The frequent parabolic use of natural phenomena may warrant the conclusion that each creative 'day' was a period of time marked off by a beginning and ending."[13]

An ancient world is also defended by conservative Bible scholar A.H. Strong in his *Systematic Theology*.[14] And in his *Encyclopedia of Bible Difficulties*, Old Testament scholar Gleason Archer points out that "it would seem to border on sheer irrationality to insist that all of Adam's experiences in Genesis 2:15-22 could have been crowded into the last hour or two of a literal twenty-four-hour day."[15]

This Christian tradition would demonstrate an open attitude toward the big bang theory—not putting it before the Bible, and not necessarily claiming that it is the only theory that could ever fit the facts—but, for the following reasons, accepting it as a reasonable theory that accords well with the bits of information that the Bible gives us about the beginning of the universe.

The Harmony of Modern Cosmology with the Bible Itself

The Bible's Harmony with an Old Universe

The theological theory of a recent creation rests, in great measure, upon a single word used in the first chapter of Genesis—the word "day," a word whose intended meaning is open to dispute. Each category of creative acts is described as taking place in one day. The fact that the word *can* refer to a great period of time rather than a solar day is clear from its nearby use in Genesis 2:4, which speaks of the "day" that all this creation activity took place. Here the Hebrew word *yom* is intended to mean a general period of time, since it includes all the "days" of creation. Most versions do not even translate the word as "day," so that it will not be confused with a solar day. The New International Version simply states "when they were created," rather than "in the day that the LORD God made." But in the original Hebrew text, the word "day" is there as clearly as it is in each of the "days" of creation.

Even solar-day proponent Henry Morris admits, "There is no doubt that *yom* can be used to express time in a general sense."[16] He goes on to mention that the word is translated as "time" 65 times in the King James Version, but as "day" almost 1,200 times, and draws conclusions from this fact. However, later he confesses that many of those 1,200 cases also refer to a more general period of time, even though they are translated as "day."[17]

Gleason Archer, an authority on ancient Semitic languages, points out that, in the Hebrew, Genesis 1 omits the definite article before each of the creation days. Rather than saying "the first day," it literally reads, "day one." Says Archer, "In Hebrew prose of this genre, the definite article was generally used where the noun was intended to be definite; only in poetic style could it be omitted."[18] This would also lead us to believe that a figurative meaning for the word *day* is intended.

The author of the Genesis creation account may have had a 24-hour day in mind, or he may have used the 24-hour day as a fitting picture to describe the stages of creation. Scripture is filled with examples showing use of this word as a figure for longer time periods, as in "the day of the Lord," "in that day," etc.

It is also possible that the author was not thinking so much of ages but of the literary structure of his narrative. Today a growing number of conservative Bible scholars hold to this position, since the writer casts the creation days within a parallel framework. The order and symmetry of this structure stands in stark contrast to other ancient creation accounts, where the world is created during violent conflicts between gods. In the Bible, however, care and design are stressed even in the literary structure.

Charting the days reveals the symmetrical beauty of days that correspond to one another both horizontally and vertically.

Days 1–3 correspond to one another vertically, each speaking of formation, while days 4–6 also correspond to one another vertically, each speaking of filling. Moreover, placing these two lists side by side reveals that each day of formation corresponds horizontally to a day of filling: day 1 corresponds to day 4 (light), day 2 corresponds to day 5 (sea and sea creatures), and day 3 corresponds to day 6 (land and land creatures). It is also possible to accept this overall framework as an artful use of a real sequence of creative events, since the geologic record's agreement with Genesis 1, in my view, is difficult to explain by coincidence alone. Volume 3 deals with this subject in great detail.

Whether one accepts the day-age interpretation or the literary framework hypothesis (or some combination of both), the inerrancy of Scripture is not violated. The Scriptures allow for the geologic ages that science reveals. Some would even say that the Bible's most *natural* interpretation necessitates long periods of time before the creation of humans.

Notice, for instance, the use of the word "generations," again in Genesis 2:4 (usually translated "account" in modern versions). The verse literally translates as "These are the generations of the heavens and the earth when they were created, in the day of their making." In Hebrew the word "generation" (*toledoth*) means the number of years between the birth of parents and the birth of their offspring, or a period arbitrarily longer.[19] The fact that it is used in the plural obviously suggests a long period of time for the creation of the heavens and the earth; it is difficult to reconcile these "generations" with the idea of only six solar days.

The rest of the Bible certainly gives us the general sense that creation was a time-consuming, monumental process, involving ages (Proverbs 8:22-31, Psalm 104, Micah 6:3, Habakkuk 3:6). Genesis, the book of beginnings, only devotes one chapter out of fifty to describe creation, and the rest of the Bible is far too general to satisfy the curiosity of a scientist. Why doesn't the Bible give us more specifics? Why, after all, doesn't Genesis tell us about each phylum and family and give us the specifics of all the processes God used to bring them about, along with their exact dates?

Obviously these facts, interesting as they would be to scientists, are not the reason for God's communication. They would not satisfy the greatest need of the human heart, the need to know not just *when* or *how* the universe was created, but *Who* created it. It is not the Bible's intention to teach science. Like the teacher/king in Ecclesiastes, we have to work for our knowledge of God's natural order. But His gift through His Word is the revelation of who He is, which we would know only imperfectly apart from special, supernatural revelation. Supernatural revelation tells us *Who* created and *why*. Natural revelation tells us something about *how* and *when*.

God's purpose through Genesis 1, for readers today as well as in Moses' time, is certainly to magnify Himself as sole Creator and to let it be known that all creation serves Him and His purposes. Nature is not to be worshipped. Only the Creator is worthy of that honor. Bible-believing science philosopher Bernard Ramm writes: "This is more effectively brought out by an absence of reference to all secondary causes. *God speaks and it comes to pass!* Expositors have been mistaken in assuming that (i) this cannot involve time, and (ii) this cannot involve process."[20]

The Bible's Harmony with Our Findings From Nature

Some people, however, feel that the Bible must be the only source of revelation for *all* truth. This single revelation position leads to the view that science is of the devil, that everything we need to know is in God's Word and to try to gain knowledge from other sources is sin. Such a position runs contrary to Scripture itself, since Psalm 19:1-4 and Romans 1:19-20 clearly tell us that God also speaks to us through creation. Above the entrance to the old Cavendish Laboratory in London is engraved Psalm 111:2: "Great are the works of the LORD; they are studied by all who delight in them."

Even though it is not the Bible's purpose to teach science, its brief account of creation in Genesis 1, sketchy and unsatisfying as it is from a scientist's viewpoint, is the only ancient text that accurately describes certain geologic truths. As we'll see in Volume 3, the theologians who advocate a young earth view must confront another difficult question: If Genesis cannot allow for a creation that took place over billions of years, why does the account fit the general descriptions of the geologic ages so well?

Both the Bible and geology testify to a progressive creation: the world was not created in an instant with everything in its place. Both tell us that the earth was once "formless and empty," and that this condition was followed by a "primitive universal ocean,"[21] which in turn was followed by the appearance of dry land (or as Eric Lerner puts it, "a gradual retreat of shallow seas from all the continents"[22]). Both agree that darkness covered the earth in its earliest history (required by the early opaque debris cloud in all theories of planet formation). Both agree that animal life first inhabited the sea. Both agree that plant life preceded land animals, and that birds preceded mammals. And now (after 150 years of studying the fossil record), both are even beginning to agree that each life form appeared abruptly, with no transitional forms between them (see Volume 2). Both testify that mammals, and finally humans, were the last to appear. Other ancient creation accounts, filled with accounts of battles and sexual encounters between mythical gods of sea, sky, and earth, explain creation in very different terms. These match neither the order nor the substance of events, according to science.

We have already discussed how the Bible harmonizes with the theory of relativity ("With the Lord a day is like a thousand years, and a thousand years are like a day") and with the laws of thermodynamics ("the heavens will all wear out like a garment"). Like the Bible, both of these fundamentals of science clearly point to a creation event. The current scientific view that this beginning must have taken place between about ten and twenty billion years ago is a bigger embarrassment for science than for theology, since this simply does not give the processes of random selection anywhere near the amount of time they would require to produce the purposeful complexity that is life. We will look further into this in Volume 2.

Die-hard recent creationists find themselves on the same side as the atheists who continue to reject the big bang's evidence for philosophical reasons. To some dissenting scientists and science writers, the big bang is especially unattractive because of its close correlation with biblical creation. When science writer John Boslough criticized big bang cosmology, he called it "the scientific model of Genesis."[23] He denounced it for its inability to account for the lack of ripples in the microwave background radiation (since he was writing in 1992, just before their discovery); but notice how he phrased his concluding criticism:

> For the time being, the big bang remains a scientific paradigm wrapped inside a metaphor for *biblical genesis*, a compelling although simplistic pseudoscientific creation myth embodying a *Judeo-Christian* tradition of linear time that led to Western ideas about cultural and scientific progress and which ordained an absolute beginning.[24]

Unlike other ancient religious texts, the Bible teaches that God is transcendent and that the universe is not eternal. In this century science has come to agree. The very existence of certain atoms shows that the elements themselves cannot be infinite in age. Science historian Owen Gingerich explains: "If the atoms were infinitely old, then radioactive uranium and thorium would have turned to lead. Their very existence tells us that they were formed at a finite time past."[25]

Even recent creationists accept the fact that the universe is expanding, but they say that God created an already-expanding universe, rather than a big bang, for the sake of the universe's stability.[26] If the galaxies were simply created without this outward motion, gravity would bring all the galaxies crashing together into a "big crunch." Either way (big bang or creation in an already-expanding mode), God's care and wisdom are shown in this expansion that establishes a stable universe.

Does the Bible even hint at an expanding universe? At least one astronomer has recognized a possible correlation. In Steven Beyer's *The Star Guide,* Beyer notes an interesting fact about the word used in the Hebrew Bible to describe the heavens in the context of creation (e.g., Genesis 1 and Psalm 19). Long before scientists had proposed any theory of an expanding universe, Hebrew lexicographers (in 1762) described this word "raquia" as expressing "motion of different parts of the same thing, at the same time, one part the one way and the other, the other way, with force."[27] In 1821, John Reid translated the word as "expansion, the heavens, from their being stretched forth."[28] Young's concordance defines the Hebrew noun *raquia* as "expanse" and the verb *raqa* as "to spread out or over." Strong's concordance defines *raquia* as "an expanse, i.e. the firmament" and *raqa* as "to pound" or "to expand."

To describe God's activity in stretching out the heavens, the Hebrews used both the verbs *raqa* and *nata,* as in the many cases in which the Bible says that God "stretched out" or "spread out the heavens" (e.g., see Job 26:7, Isaiah 40:22, 42:5, 45:12, Jeremiah 12:12, Zechariah 12:1). Of course, we have no way of knowing whether these words were purposely used to describe the expansion of the created heavens. But once again, unlike any other ancient creation account, this one certainly does not contradict the lessons science has recently taught us about the origin of our universe.

The Big Bang's Harmony with Our Witness

For Christians, talking about the stability produced by an expanding universe and the tremendous time and care God took to prepare the world for us can be a natural way to bring God into a conversation. However, advancing the young earth view poses a serious stumbling block to many who fear they must subscribe to it in order to believe the Bible. Reasonable people who have some knowledge of science will tend to dismiss our gospel along with our geology. How much better it is to present facts that witness rather than private interpretations that scare unbelievers away.

Conclusion—
The Story from Science: "In the Beginning . . ."

The conclusions drawn from classical thermodynamics, from general relativity, and from the observations of astronomy all unite to tell the same story. This century has seen the convergence of these three fields, resulting in what is probably the greatest discovery of modern science: the finding that the universe must have had a beginning—a beginning that was highly ordered and purposeful.

British theorist Edward Milne concluded his mathematical treatise on relativity with the statement: "As to the cause of the Universe, in the context of expansion, that is left for the reader to insert, but our picture is incomplete without Him."[29]

Shocking Summary Statements and Stimulating Conversation Starters

There is a conflict between modern science and some traditional beliefs among Christians, but there's no conflict with the Bible itself.

Recent creationists need to think about what they are seeing when they look up into a star-filled night. If the light from many of these stars and galaxies is coming to us from millions of light-years away (as recent creationists admit), and if God created the heavens and the earth just ten thousand years ago, then how do we see this light that took millions of years to get here? When we see a supernova in a distant galaxy, is this an event that never really happened? Is God sending us false stellar reports by light?

Is the ancient universe view only held by liberal Christians who wish to accommodate the discoveries of modern science? Actually, pre-modern Bible believers such as Josephus, Philo, Augustine, Irenaeus, Origen, Basil, and Thomas Aquinas all held that the creation "days" were not necessarily meant to be taken as literal, solar days. In this century, the ancient universe position has been held by such conservative Bible scholars as C.I. Scofield, A.H. Strong, and Gleason Archer.

The writer of the Genesis creation account may have had a solar day in mind, or he may have used the solar day as a fitting picture to describe the stages of creation. Either interpretation is legitimate; so don't let anybody bully you into thinking it's his way or no way. Anybody who feels he has to bully you in order to convince you shouldn't be taken too seriously anyway.[30]

Old and young earth creationists generally agree that the universe is expanding—they just differ about how far back we can follow the process in reverse until we come to the moment of creation. Leaders in both groups agree that this outward motion provides stability; without it gravity would act to pull all the galaxies together. Thus, whether God created the universe in a big bang or in an already-expanding mode, God's wisdom and care are displayed by astronomical observations.

Notes for Chapter 8

1. Edwin Tenney Brewster, *Creation: A History of Non-Evolutionary Theories* (Indianapolis: The Bobbs-Merrill Company, 1927), p. 109.

2. Paul M. Steidl, *The Earth, the Stars, and the Bible* (Grand Rapids: Baker Book House: 1979), p. 223.

3. S.J. Goldstein, J.D. Trasco, and T.J. Ogburn III, "On the Velocity of Light Three Centuries Ago," *Astronomical Journal* 78 (1973), pp. 122-125. Cited by Hugh Ross, *Creation and Time* (Colorado Springs, CO: NavPress, 1994), p. 98. For additional information, see Ross's book.

4. John Peacock, "Fresh Light on Dark Ages," *Nature* 355 (1992), p. 203. Cited by Hugh Ross, *Creation and Time* (Colorado Springs, CO: NavPress, 1994), p. 98. For additional information, see Ross's book.

5. Barrow and Silk, pp. 11-12.

6. Donald B. DeYoung, *Astronomy and the Bible: Questions and Answers* (Grand Rapids: Baker Book House, 1989), p.91. For another example, see Paul S. Taylor, *The Illustrated Origins Answer Book* (Mesa, AZ: Films for Christ Association, 1989), p. 6.

7. Steidl, p. 222.

8. Albert Einstein, quoted by Richard Morris, *The Fate of the Universe* (New York: Playboy Press, 1982), p. 171.

9. Bernard Ramm, *The Christian View of Science and Scripture* (London: The Paternoster Press, 1965), p. 134.

10. Ibid., p. 145.

11. Hugh Ross, *The Fingerprint of God*, second edition, (Orange, CA: Promise Publishing Co., 1989, 1991), p. 141.

12. Ibid. Also see Hugh Ross, *Creation and Time—A Biblical and Scientific Perspective on the Creation-Date Controversy* (Colorado Springs, CO: NavPress, 1994), pp. 16-17.

13. C.I. Scofield, *Reference Bible* (New York: Oxford University Press, 1917), p. 3, footnote no. 2, and p. 4, footnote no. 2.

14. Augustus Hopkins Strong, *Systematic Theology* (Philadelphia: The Judson Press, 1945), vol. 2, pp. 393-396.

15. Gleason Archer, *Encyclopedia of Bible Difficulties* (Grand Rapids: Zondervan, 1982), p. 60.

16. Henry Morris, *Scientific Creationism* (El Cajon, CA: Master Books, 1991), p. 223.

17. Ibid.

18. Archer, *Encyclopedia of Bible Difficulties,* pp. 60-61.

19. R. Laird Harris, Gleason L. Archer, and Bruce K. Waltke, *Theological Wordbook of the Old Testament,* Volume I (Chicago: Moody Press, 1980), pp. 378-379.

20. Ramm, p. 150.

21. J.W. Dawson, *The Origin of the World According to Revelation and Science* (New York: Harper Brothers, 1877), pp. 343-357.

22. Eric Lerner, *The Big Bang Never Happened—A Startling Refutation of the Dominant Theory of the Origin of the Universe* (New York: Random House, 1991), p. 400.

23. John Boslough, *Masters of Time—Cosmology at the End of Innocence* (New York: Addison-Wesley Publishing Company, 1992), p. 56.

24. Ibid. p. 223. Emphasis added. NASA's John Mather tells me that Boslough no longer criticizes the big bang theory, and at the time of this writing, Boslough is writing a book with Dr. Mather about the COBE findings.

25. Owen Gingerich, "Let There Be Light: Modern Cosmogony and Biblical Creation, *The World Treasury of Physics, Astronomy, and Mathematics,* ed. by Timothy Ferris (Boston: Little, Brown and Company, 1991), p. 382.

26. Gary Parker, "Truths That Transform" radio program, aired November 16, 1992 (Ft. Lauderdale, FL: Coral Ridge Ministries); also available on audiocassette tape.

27. Steven L. Beyer, *The Star Guide—A Unique System for Identifying the Brightest Stars in the Night Sky* (Boston: Little, Brown and Company, 1986), p. 149.

28. Ibid.

29. Edward Milne, cited by Robert Jastrow, *God and the Astronomers,* second edition (New York & London: W.W. Norton & Company, 1992), p. 104.

30. An example of such "my-way-or-no-way" bullying is found at the end of Henry Morris's *Scientific Creationism,* where Morris tells his readers: "There seems to be no possible way to avoid the conclusion that, if the Bible and Christianity are true at all, the geological ages must be rejected altogether." See Henry M. Morris, *Scientific Creationism* (El Cajon, CA: Master Books, 1974, 1991), p. 255. As we will see in Volume 3, many devout Christians, having looked carefully into the Bible and the geologic record, fully accept the truth of both. In fact, many biblical passages are most naturally understood in light of lengthy geologic ages.

PART IV

Evidence of Divine Design

How many are your works, O LORD!
In wisdom you made them all.
—*Psalm 104:24*

Come and see what God has done,
how awesome his works in man's behalf!
—*Psalm 66:5*

Question:

Does science provide evidence that the universe was *designed*, or can the apparent design be explained as well by chance?

The Whirlpool Galaxy, M51, gives us an excellent face-on view of a spiral galaxy. It is 13 million light-years distant, meaning that we are looking back in time and seeing the galaxy as it appeared 13 million years ago. —*Courtesy of Lick Observatory, University of California*

CHAPTER 9

Evidence of Design

He who has not contemplated the mind of nature
which is said to exist in the stars . . . is not able to give
a reason of such things as have a reason.

—Plato, Laws XII

A Skeptic's Questions:

The more we discover about the universe's laws, the more we learn that there are simple, natural explanations for the way things work in nature. So why do you insist that science points to a Grand Designer?

A Bible Believer's Response:

Actually, the more we discover about how the universe works, the more we understand that our universe's laws are set within very narrow, very critical parameters. Without a combination of the wildest possible coincidences, a universe capable of sustaining intelligent life would be impossible. Even unbelieving physicists have come to agree with the Bible's assertion that our universe has been very precisely prepared for *us;* they routinely tell us that the universe's conditions were very carefully "chosen," "adjusted," "fine-tuned," or as physicist Freeman Dyson says, that the universe "in some sense must have known we were coming."[1]

Addressing the "Blind Chance" Position

"Natural Law" Begs the Question

Many who take the "blind chance" position argue that any seeming purpose or order we observe in the universe can always be explained in terms of natural laws. To those who hold this view we can first point out that the idea of appealing to natural laws only begs the question. Where did the natural laws that bring the design come from? Unless the "blind chance" advocate can come up with a natural explanation for natural laws, he can give no reason for anyone to believe that the order we observe in the universe could have been brought about by chance.

The Scientific View of Natural Law—A Dialog

Chance: I'm not denying that there is beauty in nature. But the beauty or apparent design is perfectly explained by random processes.

Design: *Random* processes? Then science is helpless in predicting planetary orbits and chemical reactions?

Chance: Of course not. Science can predict those things because of natural laws. Any seeming purpose or order we observe in the universe can always be explained in terms of the laws of nature.

Design: But who *designed* the laws of nature?

Chance: No one had to design them. They're a natural part of the universe, as the term "natural law" suggests.

Design: Do you mean to suggest that the natural laws themselves run the universe?

Chance: That's right. Science can make predictions about planetary orbits and chemical reactions because the laws of nature govern the world, not some Grand Designer. That's the scientific view.

Design: It's the superstitious view, unless you can give a rational explanation for why nature should have such laws, some *cause* for their otherwise mysterious power. After all, what *are* the laws of nature? Aren't they merely our fallible attempts to describe the patterns we observe?

Chance: I suppose so.

Design: And isn't science constantly amending them? The laws of nature can't "govern" anything. This is an all-too-common error, found in introductions to science textbooks and in the speeches of science popularizers. The domain of science includes behavior, but not governance.[2] The truly scientific view is that science can't tell *what* governs the universe.

Chance: All right, then it's my metaphysical opinion and it's based on scientific principles.

Design: It's an opinion that runs contrary to the spirit of modern science. When science finally turned from the mystical notion that Nature (with a capital N) was divine[3] to the biblical concept that it was merely a divine *work*, the way was fully opened to experiment with nature without fear of sacrilege.[4] The scientific revolution exposed nature as powerless in itself. Part of turning from belief in the occult powers of trees and amulets and broomsticks was a recognition that physical objects in themselves *have* no power. So the laws that govern nature could never come from some mystic power in nature itself.

Chance: Isaac Asimov said, and I quote, "the scientific view sees the Universe as following its own rules blindly, without either interference or direction."[5]

Design: Though the science fiction writer you quote was welcome to his opinion, he misrepresented science when he called it "the scientific view," since, I must repeat, the domain of science includes behavior, but not governance. But if you want to look at the personal opinions of scientists who have been closest to the *evidence*, those who have been the actual discoverers of nature's laws have come away convinced that these laws were too perfectly contrived toward higher ends to allow us to attribute them to chance.

Chance: Which scientists came away from their discoveries convinced of that?

Design: Albert Einstein wrote that the harmony of natural law "reveals an *intelligence* of such superiority that, compared with it, all the systematic thinking and acting of human beings is an utterly insignificant reflection."[6] Robert Boyle, the father of modern chemistry, spoke of the "laws of motion prescribed by the author of things."[7] René Descartes, who *created* our modern concept of "natural law," wrote of the "laws which God has put into nature."[8]

It was partly because Michael Faraday believed that *God* was the unifying source behind the design of all phenomena that he discovered the fundamental relations between light and magnetism, and he came up with the concepts that laid the groundwork for modern electromagnetic field theory.[9]

Johannes Kepler, discoverer of the laws of planetary motion, wrote that in finding these natural laws he was merely "thinking God's thoughts after Him."[10] Francis Bacon, father of the scientific method, and Isaac Newton both went so far as to conclude that the divine Lawgiver could "vary the laws of nature" at His will.[11] Newton said that the solar system was "not explicable by mere natural causes," and that its structure could only be attributed to "the counsel or contrivance of a voluntary agent."[12]

Chance: Let me put it this way: Whenever I find a pattern of behavior that is so orderly that it becomes predictable, I have eliminated the need for divine involvement. It is explained perfectly by natural law.

Design: Let *me* put it *this* way: you admit to observing behavior which can be expressed in predictable, orderly laws of nature, not random-like laws of nature—which raises the question, "Why *are* there these orderly laws of nature? Why not chaos?" I have a supernatural explanation. Do you have a natural one?

The Need to Show Hard Evidence

The above dialog ends in an impasse—although, if the "chance" advocate had honestly answered that last question, he would have had to admit that he had no natural explanation for the orderly laws of nature. But even if the chance position can't prove that these laws exist by chance, the design position hasn't yet proved they were designed by God. The design position *has* begun to make a good argument from authority, i.e., the opinions of Bacon, Kepler, Descartes, Boyle, Newton, Faraday, and Einstein. But the skeptic still has every right to ask, "What observational evidence do you have to show that God has anything to do with the laws of physics?"

From Nothing to Nature

What we are about to explore poses an even greater problem for non-theists than the causal argument. We are no longer talking about the difficulty of explaining how something could come out of nothing, but how *nature* could come out of nothing, with all its highly specific, organized and purposeful accomplishments.

Nature's laws are anything but arbitrary. This century's observations have uncovered patterns in nature which are so precisely fixed, within parameters so narrow, that the many structural accomplishments of these patterns in our universe cannot be explained by even the wildest coincidence.

These accomplishments require a conscious, super-intelligent mind—with a particular concern for human beings. Each of the following provides good evidence of "unnatural" organization. That is, scientists have learned nothing from nature to tell us why just these required conditions should be met. We know of nothing natural about such highly specific selections.

Rather, specified complexity points to intelligent design, as scientists who seek out extraterrestrial intelligence know very well. When seeking a signal from an intelligence from space, scientists involved in the SETI (Search for Extraterrestrial Intelligence), SERENDIP (Search for Extraterrestrial Radio Emissions from Nearby Developed Intelligent Populations), and META (Megachannel Extraterrestrial Array) projects distinguish between blind chance patterns and intelligent information on the basis of highly specified selections—the same kind of purposeful selections we are about to explore.

Part of the SETI Institute's Targeted Search System (TSS) for Project Phoenix. From left to right, the racks contain: the Radio Frequency Signal Generator, Pulse Detector, Multi-Channel Spectrum Analyzer, and CW (continuous wave) Detector. —*Courtesy of SETI Institute*

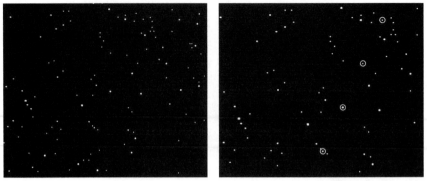

Left: Searching for triplets: While the CW Detector searches for continuous waves, the Pulse Detector searches for pulses of energy. Time increases downward and frequency to the left. Can you find at least three dots that are evenly spaced in time? Right: SETI's computer found these five pulses (circled) in less than 1/1000 of a second. New computers are scheduled to scan more frequencies in less time.
—*Courtesy of SETI Institute*

Unnatural Selection No. 1:*

The Existence of Elements Necessary for Life

The existence of the chemical elements necessary for life cannot be explained by random events or blind forces. Astrophysicist Sir Fred Hoyle is credited with the discovery of the resonances of carbon and oxygen atoms. Working with William Fowler, Hoyle discovered that, by all rights, the carbon atom, which seems to have been uniquely designed to make life possible, should either not exist or be exceedingly rare. In order to form, the carbon atom needed first to have a very precise level of a nuclear property called "resonance."** Steven Weinberg writes:

> Life as we know it would be impossible if any one of several physical quantities had slightly different values. The best known of these quantities is the energy of one of the excited states of the carbon 12 nucleus. There is an essential step in the chain of nuclear reactions that build up heavy elements in stars.[13]

---

*Viewed from a biblical perspective, none of these carefully chosen conditions I call "unnatural" selections are truly unnatural. God's selections *became* nature, and so, in a sense, natural. But the carefully chosen conditions listed here are not natural selections in a Darwinian sense; they are not explicable by any random process or by a nature without Supernatural direction. In fact, according to the biblical view, nature is completely dependent upon God, Who not only brought the universe out of nothing (Hebrews 11:2), but continues to hold all things together (Colossians 1:17).

**Resonance describes the behavior or the amount of "excitement" in the nucleus of an atom. Nuclei are normally in what is called the ground energy state, configured for stability and minimum energy. But nuclei can become excited as a result of colliding with other nuclei. When this happens, one of the protons or neutrons of the nucleus moves into a higher orbit, in much the same way that whole *atoms* are said to become excited when an electron moves from one orbit to another. Fred Hoyle correctly predicted the precise resonance that would allow carbon to form in abundance, as required by all living things.

The essential step begins with the collision of one helium nucleus with another in a star, producing a fleeting form of an element called beryllium. Robert Kirshner, chair of Harvard's astronomy department, explains what happens next:

> Amazingly enough, another helium nucleus collides with this short-lived target, leading to the formation of carbon. The process would seem about like crossing a stream by stepping fleetingly on a log. A delicate match between the energies of helium, the unstable beryllium and the resulting carbon allows the last to be created. Without this process, we would not be here.[14]

This improbable step allows not only for the formation of carbon, but of all the elements that are built up in subsequent nuclear reactions: oxygen, nitrogen, and the other heavy elements required for life. Without this step, the universe would consist almost entirely of hydrogen and helium. Thus, everything hinges on the resonance (energy level) of the carbon nucleus. A resonance slightly below a certain energy level would not allow carbon to form. An energy level just a little higher than the actual resonance of carbon would instantly destroy it.

Before physicists found the resonance of carbon, Fred Hoyle precisely predicted what it should be—based solely on life's need for it. "He hit it virtually on the money," physicist James Truran told me, "because when they finally went back and did a further experimental study on the carbon, they found that this particular resonance, this particular excited state structure, was right where he had predicted it should be."[15] When Hoyle then calculated the chances that such resonances should exist by *chance* in these elements, he said that his atheism was greatly shaken.[16] In explanation, Hoyle wrote the following in the November 1981 issue of *Engineering and Science*:

> A common sense interpretation of the facts suggests that a superintellect has monkeyed with physics, as well as with chemistry and biology, and that there are no blind forces worth speaking about in nature. The numbers one calculates from the facts seem to me so overwhelming as to put this conclusion almost beyond question.[17]

Princeton physicist Freeman Dyson writes of other "lucky accidents" in the atomic structures necessary for life: "Without such accidents, water could not exist as a liquid, chains of carbon atoms could not form complex organic molecules, and hydrogen atoms could not form breakable bridges between molecules."[18]

The atomic structure coincidences led physicist Richard Morris to speculate about conscious planning: "How is it that common elements such as carbon, nitrogen, and oxygen happened to have just the kind of atomic structure that they needed to combine to make the molecules upon which life depends? It is almost as though the universe had been consciously designed . . . ."[19]

Even Carl Sagan says, "the origin of life on suitable planets seems written into the chemistry of the universe."[20] What Sagan never explains, of course, is *how* life got written into the chemistry of the universe. Moreover, Sagan acknowledges that the laws of nature cannot be randomly shuffled: "It is easy to see that only a very restricted range of laws of nature are consistent with galaxies and stars, planets, life and intelligence."[21]

Mother Earth and the Gaia Hypothesis

Though this hypothesis may not be worthy of including as an "unnatural selection," it does illustrate the fact that those who study nature are often forced to conclude that there must be something more than accidents behind it. To explain how the earth maintains life, many scientists adhere to the Gaia (pronounced "gay-yuh") hypothesis, the view that the earth's maintenance of the required temperature range and oxygen levels is no accident. The earth seems to have been designed to regulate itself; otherwise, temperatures and oxygen levels on earth have inexplicably remained within the narrow parameters required for life for long periods of time.

Scientist James Lovelock points out that oxygen levels should have been fluctuating wildly over recent ages.[22] But: "Were oxygen levels to rise above 25%, spontaneous fires would break out; if they dropped below 15%, many higher life-forms would suffocate."[23] Thus we know that oxygen has remained close to 21% of the atmosphere on this planet for at least as long as mammals have been around. Other factors involve the creation of a protecting ozone layer, the ocean and the hydrologic cycle (which brings water to almost every region of the planet).

Of course, the contention that Mother Earth should possess such self-regulating powers is itself a subtle appeal to a higher power (the word *Gaia* is the name of the ancient Greek goddess of the earth). Some, such as Tyler Volk of New York University and David Schwartzman of Howard University, advocate the Gaian contention that life itself regulates the atmosphere's temperature and composition in order to suit it. Others, mainly geochemists, assert that inorganic processes better account for the finely-adjusted conditions that life has enjoyed for so long on this planet—

though life itself may provide a "weak climatic stabilizing mechanism."[24] Either way, current scientific theories illustrate the need to explain perfectly adjusted conditions.

One explanation for all these "lucky accidents" and "coincidences" that resulted in a habitable world is found in Isaiah 45:18: "For this is what the LORD says—He who created the heavens, He is God; He who fashioned and made the earth, He founded it; He did not create it to be empty, but formed it to be inhabited."

Earthrise from the moon. Apollo 10 astronauts took this picture while orbiting the moon, about 100,000 miles from earth. —*Courtesy of NASA*

Unnatural Selection No. 2:

The Ratio of Proton to Electron Mass

Physicists tell us that the proton is 1,836 times heavier than the electron. Why? We know of no natural reason. We only know that if it were much different, the required molecules would not form, and there would be no chemistry, no life, and no physicists to wonder about it.

Theoretical physicist Stephen Hawking mentions this ratio between the masses of the proton and the electron as one of many fundamental numbers in nature, and he says, "The remarkable fact is that the values of these numbers seem to have been very finely adjusted to make possible the development of life."[25]

Unnatural Selection No. 3:

The Relative Strength of Nature's Four Fundamental Forces

The strength of each of nature's fundamental forces seems to have been carefully selected to bring about just the kind of universe that can sustain life. Physicists speak of four fundamental forces in nature: gravity, electromagnetism, and the strong and weak nuclear forces. If any one of them had a slightly different strength, we'd have a universe in which life would not be possible. Either we'd have no atoms, or we'd have atoms but no stars or planets. Physicist Richard Morris writes:

> Every one of these forces must have just the right strength if there is to be any possibility of life. For example, if electrical forces were much stronger than they are, then no element heavier than hydrogen could form. . . . But electrical repulsion cannot be too weak. If it were, protons would combine too easily, and the sun . . . (assuming that it had somehow managed to exist up to now) would explode like a thermonuclear bomb.[26]

Concerning the strong nuclear force, astronomer Hugh Ross says, "If the strong nuclear force were slightly weaker, multi-proton nuclei would not hold together. Hydrogen would be the only element in the universe."[27] Richard Morris explains the other side: "Stronger forces would cause all of the primordial hydrogen—not just 25 percent of it—to be synthesized into helium early in the history of the universe. And without hydrogen, the stars could never begin to shine."[28]

In his book, *Disturbing the Universe*, physicist Freeman Dyson adds:

> A similar but independent numerical accident appears in connection with the weak interaction by which hydrogen actually burns in the sun. The weak interaction is millions of times weaker than the nuclear force. It is just weak enough so that the hydrogen in the sun burns at a slow and steady rate. If the weak interaction were much stronger or much weaker, any forms of life dependent on sunlike stars would again be in difficulties.[29]

As another example of his "very finely adjusted" numerical values in nature, Stephen Hawking points out that "if the electric charge of the electron had been only slightly different, stars either would have been unable to burn hydrogen and helium or else they would not have exploded,"[30] meaning that the heavier elements necessary for life would not have been available. He concludes, ". . . it seems clear that there are relatively few ranges of values for the numbers that would allow the development of any form of intelligent life."[31]

Unnatural Selection No. 4:

Protein Formation

Fred Hoyle and his colleague Chandra Wickramasinghe calculated the odds that all the functional proteins necessary for life might form in one place by random events. They came up with a figure of one chance in $10^{40,000}$ (that's 1 with 40,000 zeros after it). Since there are only about 10^{80} atoms in the entire universe, the physicists concluded that this was "an outrageously small probability that could not be faced even if the whole universe consisted of organic soup."[32]

Because the facts cannot be explained by random processes here on earth, and since Hoyle will not accept a biblical explanation, he has aligned himself with the panspermia position, the hypothesis that the seeds for life must have come from other regions of the universe. Hoyle concludes:

> Life could not have originated here on the Earth. Nor does it look as though biological evolution can be explained from within an earthbound theory of life. Genes from outside the Earth are needed to drive the evolutionary process. This much can be consolidated by strictly scientific means, by experiment, observation and calculation.[33]

Fred Hoyle, Thomas Gold, Leslie Orgel, Swedish physicist Svante Arrhenius (who first coined the word "panspermia," meaning "seeds everywhere"), and Nobel prize-winner Francis Crick (who first broke the DNA code) hold to varying versions of panspermia. All have concluded that natural processes alone cannot explain the specified complexities they have studied, whether in protein formation or in the encoded information of DNA. All have been forced by their own discoveries to conclude that life on Earth is far too improbable to appear here without a lot of help from outside.

Gold claimed that life must have been sent here in a spaceship from a dying civilization, and that perhaps just the astronauts' bacteria survived the journey. Crick and his colleague Leslie Orgel propose that just the genetic material was sent in the first place, aimed specifically at this perfect planet (directed panspermia). Arrhenius, Hoyle, and Wickramasinghe have simplified the hypothesis still further, saying that the genetic material is being sent all over the universe *without* a spaceship. The genes simply ride the stellar winds at thousands of kilometers per second, taking root wherever a planet offers the right conditions. Of course, the panspermia hypothesis only begs the question of how life originally began. If natural laws cannot explain how life began on this ideal planet, how can they explain the formation of life at *any* location?

The fact that respected scientists (among them, Nobel laureates and discoverers of nature's constants) should have to resort to speculations as wild as panspermia gives us a good indication of how difficult it is to explain the formation of life's building blocks by any earth-bound, natural process.

Unnatural Selection No. 5:

Balance Between Gravitational Force and Electromagnetic Force in Stars

Another numerical coincidence occurs in the relative strengths of the two forces that balance each other in every star: the gravitational force, which holds the star together, and the electromagnetic force, which allows the star to radiate energy. Most stars lie in the very narrow range that allows them to operate like our sun. Outside of this range they would be either red giants or blue dwarfs, making it impossible for them to sustain human life on planets.

The correlation between the strength of these two forces is very precise. According to the calculations of physicist Brandon Carter, if the strength of gravitation force were altered by a mere one part in 10^{40}, we'd have a world in which all stars would be either red dwarfs or blue giants.[34] English physicist and mathematician Paul Davies says, "Stars like the sun would not exist, nor, one might argue, would any form of life that depends on solar-type stars for its sustenance."[35]

And physicist Edward Kolb of Fermi National Accelerator Laboratory concludes: "It turns out that 'constants of nature,' such as the strength of gravity, have exactly the values that allow stars and planets to form . . . . The universe, it seems, is fine-tuned to let life and consciousness flower."[36] Science, he admits, may never be able to tell us why this should be.[37]

Unnatural Selection No. 6:

Universe Kicked Off in an Unnatural Low Entropy State

The initial low entropy state of our universe still has no natural explanation. Entropy can be defined as a measure of the amount of disordered energy in a system, that is, the amount of energy that is unavailable for work. The big bang cannot explain why the universe began with such a high potential for order rather than disorder. Paul Davies puts it this way: "If the big bang was just a random event, then the probability seems *overwhelming* (a colossal understatement) that the emerging cosmic material would be in thermodynamic equilibrium at maximum entropy with zero order."[38] We have no modern theory to explain why, contrary to the second law of thermodynamics, our universe somehow got into such an orderly state. Science has brought us no further on this question than when Isaac Newton first asked, "Whence arises all that order and beauty we see in the world?"[39] Davies explains:

> The mystery of how the universe got into its low-entropy state has exercised the imagination of several generations of physicists and cosmologists . . . . Given a random distribution of (gravitating) matter, it is overwhelmingly more probable that it will form a black hole than a star or a cloud of dispersed gas. . . . The present arrangement of matter and energy, with matter spread thinly at relatively low density, in the form of stars and gas clouds would, apparently, only result from a very special choice of initial conditions. Roger Penrose has computed the odds against the observed universe appearing by accident . . . . He estimates a figure of $10^{10^{30}}$, to one."[40]

Unnatural Selection No. 7:

Balance Between Universe's Expansion and Collapse

The expansion force of our universe is precisely balanced by the needed gravitational force: enough gravity so that matter could collect to form galaxies, and enough expansion force so that the universe doesn't come crashing back in on itself early in the expansion process.

Richard Morris says, "If our universe had been expanding at a rate that was slower by a factor of one part in a million, then the expansion would have stopped when it was only 30,000 years old, when the temperature was still 10,000 degrees."[41] Astronomers John Barrow and Joseph Silk give the other side: "Worlds expanding much faster than the critical rate would almost certainly be devoid of stars and galaxies and hence the building blocks of which living beings are made."[42] Calculations show that matter would have been moving outward too rapidly to allow any of it to gravitate together into clumps.[43]

Either way, there would be no opportunity for stars to form, no opportunity for planets, and no opportunity for life. Indeed, as one literal translation of Psalm 19:1 says, "The expansion shows His handiwork."[44]

After completing most of his work on singularities and the big bang in 1983, Stephen Hawking told a reporter:

> The odds against a universe like ours emerging out of something like the big bang are enormous. . . . I think clearly there are religious implications whenever you start to discuss the origins of the universe. There must be religious overtones. But I think most scientists prefer to shy away from the religious side of it.[45]

George Smoot on the Critical Balance Between Expansion and Collapse[46]

I mentioned before that George Smoot told me that the universe had to be made very perfectly in order for it to grow to its great size and last so many billions of years. "Otherwise," Dr. Smoot explained, "imperfections would mount up and the universe would either collapse on itself or fly apart." When he said that the density must be within critical parameters—within one percent—skeptic, that I am, I asked him how he knew that.

Heeren: Now do we actually know that? Because I understand that there's still a lot of controversy about whether it's going to expand forever or collapse back on itself. Is the difference really a matter of one percent?

Smoot: Yeah, well, it's even more than that. . . . If you look at the universe today, you can very easily realize that the density in the universe is somewhere between one percent of what's needed to cause it to collapse and twice what is needed to cause it to collapse. Now that seems like a pretty wide range—that's a factor of a hundred, right? But if you extrapolate backwards in time, you find that it gets closer and closer to being exactly the right amount, because it's an unstable situation. Even if you're only a little bit away from being exactly the right amount, it will deviate very rapidly, and either the universe will fly apart, or it will have collapsed already. That's why we know that it's so close

> now, after fifteen billion years. It's sort of like this: imagine you
> have a target. And you hit it—it's a typically sized archery tar-
> get—but the arrow was shot from *Pluto,* right? And so maybe it
> missed the target by a hundred target widths—it still is an
> impressive shot, right?

Heeren: Right.

Smoot: And that's exactly the kind of analogy that's right for how close
>   to being critical [this is]. And the fact is, we can visibly see from
>   emitted starlight at least one percent of what we need. And if it
>   was more than twice what we need it would have already col-
>   lapsed. And so we know we were very close. And so as you go
>   back, it means that we know the arrow was *very* close to the
>   exactly right trajectory for most of its flight time.

Once one takes into consideration the "flight time" of our universe, the chance that our universe's expansion rate should meet the requirements for the balance we observe becomes ridiculously small; and the problem for those who wish to explain this occurrence by naturalistic means becomes equally large. Stephen Hawking expresses the problem this way:

> Why did the universe start out with so nearly the critical rate of expan-
> sion that separates models that recollapse from those that go on expand-
> ing forever, so that even now, ten thousand million years later, it is still
> expanding at nearly the critical rate? If the rate of expansion one second
> after the big bang had been smaller by even one part in a hundred thou-
> sand million million, the universe would have recollapsed before it ever
> reached its present state.[47]

Applying the general theory of relativity to the size of the universe, it is possible to calculate the critical density (that is, the ideal mass per unit volume) that strikes a compromise between the two alternatives, between matter that is too dispersed to form galaxies on the one side and an early collapse of the universe on the other. The result shows a critical density of about 10^{-29} gram per cubic centimeter on average across the universe. This means the cosmos must contain a total number of about 10^{80} particles.

Up until the end of the 1970s, most astronomers were convinced that the mass of the universe was less than the required amount. The universe seemed destined to expand forever, ending in "the big chill." By 1980, however, as a result of new discoveries involving mass from an invisible source (called dark matter), scientists acknowledged that the universe could be closed. The expansion could eventually reverse itself, and the universe could collapse into a "big crunch."

In 1985, participants at an International Astronomical Union symposium took a vote on how closely they believed the mass of the universe matched the critical mass. In other words, is the universe open (expanding forever) or closed (eventually collapsing)? Of the fifty-nine astronomers who gave values for the mass, only two said the universe was closed. Twenty-eight gave values that were within a tenth of one percent of the critical mass. And seventy-one said they didn't know.[48] Today, the very fact that the universe has lasted as long as it has, combined with new observations (such as motions of galaxy groupings on huge scales, a billion light-years across)[49] continues to convince many physicists that the universe must contain very close to 10^{80} atoms, the apparent number required for balance.[50] For some details on how these measurements are made, see Bonus Section #2.

Einstein lectures at Mount Wilson's Hale Library in 1931. The equation he writes, "$R_{ik}=0$?," raises the question: Is space flat? R_{ik} refers to the curvature of space, which can be either positive, negative, or flat (zero). These correspond respectively to a density for our universe that is either below, above, or equal to the critical density. —*Courtesy of The Huntington Library*

Geometry of Space

Euclidean

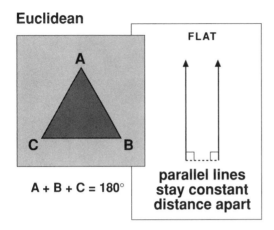

A + B + C = 180°

FLAT

**parallel lines
stay constant
distance apart**

Gauss-Lobachevski
Hyperbolic

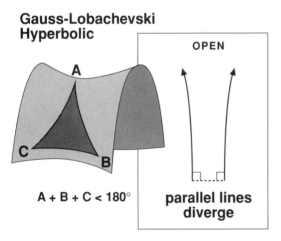

A + B + C < 180°

OPEN

**parallel lines
diverge**

Riemannian

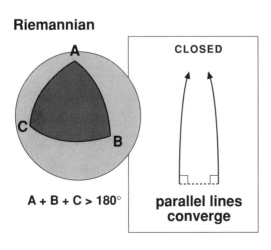

A + B + C > 180°

CLOSED

**parallel lines
converge**

If the density of the universe is precisely adjusted to the critical density, then the universe is said to be flat (top figure). Below this critical density, the universe is said to be open (center), meaning that the universe will tend to expand forever. In this case, space is said to be hyperbolic, meaning that two parallel lines will diverge. If, however, there is enough matter in the universe to pull the universe back together (bottom), then the universe is said to be closed. Parallel lines converge, meaning that a spaceship traveling in a straight line will eventually reach its starting place. If we could see far enough, we could see our own Milky Way from the other side (and we could see the backs of our heads).
*—Illustration © 1995
Christopher Slye*

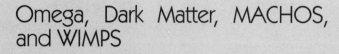

Omega, Dark Matter, MACHOS, and WIMPS

NOTE FROM CARL: AGAIN, IF YOU'RE LIKE ME AND YOU
DON'T SEE HOW DARK MATTER MATTERS, YOU MIGHT
SKIP OVER THIS SECTION.

A discussion of the critical density of our universe gives us a chance to mention a few of the hot topics in cosmology today. Astrophysicists have given a name to the ratio of the universe's actual density to the critical density: Omega. The thought that an accidental beginning of our universe should have produced an Omega of exactly 1 by a lucky accident seems exceedingly unlikely. Yet, since the 1970s, Princeton physicists Robert Dicke and Jim Peebles have argued persuasively that our universe could not be here unless Omega was 1 or *extremely* close to 1.

We already know from astronomical observations that the visible stars have enough mass to yield an Omega of almost .01 (1 percent of the mass needed for the critical density). And the fact that the universe has not yet collapsed means that there cannot be so much mass that Omega is more than 2 (twice the mass needed for the critical density). As George Smoot pointed out, if we are near Omega now, we must have been much nearer in the past. When we go back to the beginning—say three minutes after the big bang—calculations show that Omega must have been within one hundred million millionths of 1.[51] At Planck time, Omega must have been, for all practical purposes, exactly 1.

Another reason we know that Omega must be at least very near to 1 is that higher or lower figures would not allow some regions to collapse, resulting in the local groupings of galaxies and clusters that we observe today.[52] Either no local regions could collapse to form such structures, or all regions would collapse.

Again, without inflation, opposite sides of the visible universe would not have been in causal contact with one another, meaning that the critical density would not be the result of one extremely critical "accident," but of many extremely critical accidents. The inflation scenario can explain how, once this critical density was achieved at the beginning, it could continue to spread for billions of years and reach all parts of the universe. But inflation cannot explain how the density of the universe was so precisely adjusted at the beginning. It exchanges the wonder of a perfect balance in the expansion rate for the wonder of a perfect balance among physical constants at the beginning.

For Bible believers, such conditions are the mark of the Creator's faithfulness to us and to a plan that was in His mind before the beginning of our world. The Bible tells us that God created the universe with laws that would serve His purposes; He established initial conditions that would result in an enduring world: "Your faithfulness continues through all generations; you established the earth, and it endures. Your laws endure to this day, for all things serve you" (Psalm 119:90-91).

A host of strange-sounding, theoretical objects and particles have been proposed mainly because science now takes very seriously the concept of an extremely precise creation event, one that began with a critical density against enormous odds. This conviction has led to the greatest quest of cosmologists today: the search for the missing mass. Critical density means there should be a certain amount of mass in our universe, and so far scientists can account with certainty for no more than 6 percent of it.

Using the velocities of the stars in swirling galaxies, the mass of these galaxies can be calculated. Astronomers have learned that there is more to galaxies than what meets the eye: their swirling velocities cannot be explained apart from great volumes of unlit, unseen matter that must surround them. In fact, this dark matter outweighs the visible matter ten times. When astronomers estimate all the mass of all the *visible* matter in the universe, they arrive at a figure that is only about a half of a percent of the critical density required to explain our universe (Omega = .005). And when astronomers add all the *dark* matter surrounding galaxies, they still come up with a figure that is no more than 5 percent of the critical amount (Omega = .05). Hence we have a need for the strange-sounding, theoretical objects and particles that must make up the other 95 percent of our universe.

Even the less exotic 5 percent requires some explaining, since only a tenth of this is visible to us. Astronomers have thus speculated on objects they call MACHOs (massive compact halo objects). These may consist of Jupiter-like objects, burned-out stars (like white dwarfs or neutron stars), or even black holes.* And these are all composed of the kind of matter we already know about, called baryons: protons, neutrons, and associated electrons.

To account for the other 95 percent of matter required to meet the crit-

---

*Taking advantage of a phenomenon called gravitational micro-lensing, several groups now report finding some dark objects in the large "halo" region surrounding our Milky Way. Having masses about a tenth that of our sun, these appear to be tiny stars, brown dwarfs or massive planets. Such objects may prove to form a substantial amount of the dark matter in the halos of most galaxies. They may help to account for the fact that the outer parts of galaxies rotate much faster than they should. Calculations based on Newton's laws tell us that the gravitational force associated with this visible mass is too small to produce such rapid rotation.

ical density, physicists have proposed that this matter may be non-baryonic. This exotic form of matter would apparently have a gravitational influence on our universe without interacting electromagnetically; and so it would not be visible to us. Hypothetical exotic matter comes in two varieties: hot and cold. The most important thing to know about them is that hot dark matter is fast moving and dispersive; cold dark matter is slow moving and cohesive. Hence cold dark matter may attract hydrogen to itself and help clump matter together into galaxies, while particles of hot dark matter would have been streaking away near the speed of light during the early universe's epoch of galaxy formation. Cold "clumps." Hot "hurries" away.

Theories focused especially on cold dark matter in the 1980s, and a family of hot dark matter called WIMPs (weakly interacting massive particles) was popular during the early 1990s. The theories that meet the most requirements for early galaxy formation today employ a combination of both cold and hot dark matter. However, as quasars and galaxies have been observed at greater and greater distances, the time of earliest structure formation has been pushed back to earlier eras than any of our present theories can accommodate. As I write this, cosmologists are striving to solve the puzzle of how any of these hypothetical particles could have coalesced into galaxies fast enough.

The relevance of these speculations to our discussion is that they show how seriously science must take the fact that the density of our universe apparently meets, or comes extremely close to meeting, the critical density. Exotic matter is sought because of this fact. In his book, *Wrinkles in Time,* George Smoot explains why this critical density is so critical to *us:* "When we learn of the consequences of Omega being anything other than precisely 1, we see how very easily our universe might not have come into existence: The most minute deviation either side of an Omega of 1 consigns our potential universe to oblivion."[53]

Some of those who depend more strictly on observations than on theory believe the universe will expand forever, since a considerable amount of theoretical exotic matter (95 percent of our universe's mass) is required to close the universe. Here, for example, is Princeton physicist Jeremiah Ostriker's response to my question about whether he thinks the universe will continue to expand forever or eventually collapse:[54]

Ostriker: I'd say the best evidence is that it will continue to expand forever. Even though people like to have an Omega-equals-one universe, the best evidence is that it's less than one, which means that it will expand forever.

Heeren: What kind of evidence are you looking at to come up with the idea that it's expanding forever?

Ostriker: I'm looking at whatever our best estimates are for the measurements of the amount of mass in the universe—not our ideological beliefs.*

The fact that the evidence argues against a closed cosmos has implications for those who would like to believe in an oscillating universe (a universe that rebounds eternally, yielding multiple universes over vast stretches of time). In Chapter 5, I cited theoretical problems with this "re-bang hypothesis." Now we see that our best observational evidence also militates against such a concept. A universe with less than the critical mass must keep expanding, making re-bangs impossible. Despite the tremendous odds against our finely tuned universe, it appears to have had just one shot at accomplishing all that we see.

And as George Smoot pointed out, even if we miss the target (of critical mass) by a hundred target widths, if the arrow was shot from Pluto, it's still a pretty impressive shot. We still know that "the arrow was *very* close to the exactly right trajectory for most of its flight time."

Related to the ideal selection of a density for the universe is the ideal selection for the local density of stars within our galaxy. Physicist Freeman Dyson points out that the average distance between stars also meets an important requirement. He says, "The universe is built on such a scale that the average distance between stars in an average galaxy like ours is about twenty million million miles, an extravagantly large distance by human standards."[56] But he writes, "If the distances had been smaller by a factor of ten, there would have been a high probability that another star, at some time during the four billion years that the earth has existed, would have passed by the sun close enough to disrupt with its gravitational field the orbits of the planets,"[57] thus destroying life on earth.

---

*Evidently the Princeton cosmologists are more practical-minded than most of their colleagues. Another renowned Princeton physicist, Jim Peebles, has put forth the theory that our universe may contain no exotic matter whatsoever. The obvious advantage of this view is that it tries to explain the universe without resorting to anything we don't already know about. He estimates that our universe's mass (most of it dark, but not exotic) may total only about 20 percent of the critical density. This means that inflation never happened, since inflation theory requires that the critical density be met. But once again, unlike the big bang framework itself, the inflationary big bang is built upon many suppositions, not upon any hard evidence. According to Peeble's straightforward scenario, the big bang produced radiation, which produced light elements, which broke up into gas clouds, which rapidly contracted into stars and galaxies. The drawback to this scenario, in the view of those looking for purely natural explanations, is that it requires an inexplicably precise balance between sizes and numbers of fluctuations at the beginning that could eventually produce galaxies.[55]

Unnatural Selection No. 8:

Slight Excess of Matter over Antimatter

Antimatter sounds like something that was invented especially by the writers of *Star Trek*, a handy sci-fi plot device that is occasionally employed to threaten our heroes with annihilation. However, English quantum physicist Paul Dirac could not avoid the existence of "antiparticles," try as he might, when in 1929 he made calculations that combined certain elements of special relativity and quantum mechanics. He concluded that whenever energy was converted to mass, matter must be created in equal amounts of matter and antimatter. Each atomic particle must then have its opposite. Protons must have antiprotons as their antimatter counterpart. If there are electrons, then there must be anti-electrons, now called positrons. Each antiparticle has exactly the same mass, behavior, etc., as its counterpart—except that its electrical charge is opposite.

The idea of converting mass to energy is easy enough to grasp: We see it when we see the sun shine or a candle burn. In the case of candles, the amount of mass that is actually converted to energy is measured in billionths of the original mass. Nuclear fusion (as in a hydrogen bomb or the sun) converts a half of one percent of the mass into energy. But if matter and antimatter were ever to come into contact, a perfectly efficient conversion of matter into energy would result; that is, 100 percent of the matter would be converted into energy. Obviously, scientists would like to find and control such an efficient energy source, just as the *Star Trek* writers used antimatter to fuel their spaceships.

Antimatter *was* found—and the proposal on paper entered the realm of observational science—in 1932. Caltech's Carl Anderson detected a particle in his cloud chamber that had the same mass as an electron—but it was positively rather than negatively charged. He photographed its vapor trail, showing that it took the path of a positively charged particle. Such positrons are thought to be rare, probably resulting from occasional collisions with high energy cosmic particles in the upper atmosphere. Dirac and Anderson later received Nobel prizes for their work, as did the physicists in the 1950s who used a particle accelerator at the Lawrence Berkeley Laboratory to generate their own antimatter. By creating high energy collisions between protons, Emilio Segré and Owen Chamberlain produced pairs of protons and antiprotons, demonstrating that when energy is converted into matter, an equal amount of antimatter is created.

A puzzling question remained: Why didn't the original creation of matter result in equal amounts of matter and antimatter throughout the universe? Some invoked the anthropic principle, "explaining" the phenomenon simply by saying that the universe simply *must* be this way or we wouldn't be here to talk about it. If matter and antimatter had been produced in equal quantities, as Dirac's calculations showed we should expect, a tremendous annihilation would have occurred. Our universe would consist of a vast soup of photons, which, if it lasted as long as our actual universe, would mean that the microwave background radiation would be about all that there would be.

Hannes Alfvén (of plasma theory fame) proposed a more logical-sounding explanation for the puzzle: According to Alfvén, half of the universe *is* composed of antimatter, just as we would expect in a natural transformation of energy into mass. We just can't detect the fact that half of the stars out there are antistars, since antimatter produces the same appearance and spectrum as matter. Alfvén's proposal predicts a certain amount of stray antiparticles from antistars to continually bombard our upper atmosphere, where most would be destroyed. During the 1970s, George Smoot led a team of researchers in a search for these antiparticles via high altitude balloon tests. Their tests showed not a single antiparticle. These negative results were inconsistent with Alfvén's proposal of an alternative to the big bang (involving plasma and antimatter explosions).

Today scientists have a better explanation, but it has forced them to acknowledge yet another finely-tuned parameter to bring about our habitable universe. Physicists believe that in the very first moment of the big bang, matter and antimatter were both produced, but particles of matter must have slightly outnumbered particles of antimatter by a critical amount. Particles of matter and antimatter annihilated one another, leaving behind a comparatively small amount of matter (which went on to become the baryons, or material particles, of our universe) and a great amount of light energy (photons, which went on to become the cosmic background radiation). The predicted result matches what we now observe, a universe with a ratio of at least a billion photons for every baryon. But the problem for physicists remains: Why should matter have been created with a slight excess of matter over antimatter?

Nobel prize-winning physicist Steven Weinberg explains how critical this small excess of antimatter must have been:

> If there had not been a small excess of electrons over anti-electrons, and quarks over antiquarks, then ordinary particles like electrons and quarks would be virtually absent in the universe today. It is this early excess of matter over antimatter, estimated as one part in about 10^{10} [that's one part in 10 billion], that survived to form light atomic nuclei three minutes later, then after a million years to form atoms and later to be cooked to heavier elements in stars, ultimately to provide the material out of which life would arise.[58]

In the same article, Weinberg reflects on the fact that our science simply cannot explain why such initial conditions should have existed. Physicists can only speculate that either it was a tremendously lucky fluke for us, or that there may be some natural reason, as yet unknown. George Smoot, one-time hunter of antimatter, writes:

> It may appear from this description that the universe—and our existence in it—was the result of a lucky break: a slight excess of matter produced as a result of violating rules at the right time. This may be but one of countless possible outcomes of that minuscule slice of time following the big bang, or it may be an inevitability given the laws of physics that operated then. We do not yet know.[59]

Either the slight excess of antimatter somehow "violated the right rules at the right time," or it is built into the universe as a yet unknown physical constant. Either way, it appears that a very precise selection was made once again.

Unnatural Selection No. 9:

Centrifugal Force Perfectly Balances Gravitational Force

The outward force of spinning solar systems and galaxies precisely balances the inward tug of gravitation. In his *Principia*, Isaac Newton first described how this centrifugal force is precisely counterbalanced by the centripetal force of gravity. Speaking of how the moon is retained in orbit about the earth by this balance, he wrote: "If this force [of gravity] was too small, it would not sufficiently turn the moon out of a rectilinear [straight] course: if it was too great, it would turn it too much, and draw down the moon from its orbit towards the earth. It is necessary, that the force be of a just quantity . . . ."[60]

As a result of this inexplicable balance, everything in the universe that revolves is kept from crashing in on itself, from this solar system to the farthest galaxy. And everything in the universe *is* revolving, as if in a never-ending waltz.* There is no natural reason why the outward force produced by this revolving should so perfectly counterbalance gravitation, except

---

*Also, eight of the nine planets of own solar system happen to follow highly circular orbits, despite the fact that NASA must plan with great precision to put a satellite into a nearly circular orbit, and scientists find it extremely difficult to keep a satellite in such an orbit for extended periods. Modern theories of chaos predict that abnormalities should have developed in our planets' orbits long ago, producing wildly elliptical orbits. If Jupiter had fallen into such an orbit, it would bring showers of asteroids into our planet's path. Instead, Jupiter serves as a cosmic vacuum cleaner, attracting asteroids and comets (like 1994's Comet Shoemaker-Levy 9) to itself and keeping our earth's path clear for extended periods.[61]

that if it didn't, all these spinning systems would either crash in on themselves or fly apart, and once again, our universe would not be a place for intelligent beings.

Notice the number of times we have described the precise requirement for things to be "held together": the strong nuclear force has to be just right for nuclei to hold together, the weak interaction of burning hydrogen in every star must be precisely balanced with the gravitation that holds it together, centrifugal force must precisely balance gravitational force to hold solar systems and galaxies together, etc. With regard to all this holding together, both believers and unbelievers should find it interesting that Colossians 1:15 characterizes the Creator as one who "is before all things, and in Him all things hold together."

Conclusion

Many people (including many scientists) seem to have put an inordinate amount of faith in the natural laws themselves, as if an understanding of the equations and descriptions of physical laws can explain *why* we have these particular, "just-right" physical constants and not others.

Today, scientists are searching for a short formula that would express a grand unification of all of nature's laws. This quest for a "Theory of Everything" (called TOE) demonstrates that scientists have come to expect unity and simplicity at the heart of all complexity in nature. Experience with nature has taught many to expect a common source for all the many harmoniously working systems we observe. But what if scientists were to find such a formula? Would that explain how everything came to be—and came to be in a way that was so perfectly suited to us? Having considered this question, Stephen Hawking says:

> Even if there is only one possible unified theory, it is just a set of rules and equations. What is it that breathes fire into the equations and makes a universe for them to describe? The usual approach of science of constructing a mathematical model cannot answer the questions of why there should be a universe for the model to describe. Why does the universe go to all the bother of existing?[62]

Hawking's oft-quoted remark from *A Brief History of Time*, "What place then for a Creator?" needs to be put in its proper context—as a question springing from his mathematical proposal of a universe without a beginning—and its ramifications for *one particular type of God*.[63] After describing how most people see God merely as the one who winds up the universe and then lets it go, he shows how his mathematical proposal of a universe without a beginning would dispatch such a weak concept of a

Creator. Hawking himself wonders whether a universe that needs a Creator to exist should not also expect to feel His involvement in other ways: ". . . does he have any other effect on the universe?"[64] This seems to be the point he made in his remarks to me also, when he said that his no-boundary proposal does not imply God's non-existence, but it may affect our ideas about His *nature*.[65] In any case, the evidence has led Hawking to recognize a critical place for a Creator—not just as the universe's initiator, but as that which "breathes fire into the equations and makes a universe for them to describe."[66]

Like Hawking, other physicists see the logical necessity for something more than a set of physical laws at the beginning. Theoretical physicist Heinz Pagels contemplates the void before the beginning: "Where are these laws written into that void? What 'tells' the void that it is pregnant with a possible universe? It would seem that even the void is subject to law, a logic that exists prior to space and time."[67]

Difficulties in proposing a naturalistic cause for our universe's *design* inevitably bring us back again to difficulties in proposing a naturalistic cause for its *existence*. Ultimate causes are said to be outside of the domain of science; yet the questions of science inevitably point back to an ultimate cause—both of existence and of design—and science can give us no information about either.

This chapter has listed only nine "unnatural" selections out of a much longer list. Astronomer Hugh Ross lists twenty-five parameters for the universe, which he says "must have values falling within narrowly defined ranges for life of any kind to exist."[68] Princeton physicist Freeman Dyson represents many scientists who have recognized what appears as purposeful preparation, a perfect plan in all of nature's laws that exists especially for our benefit. He writes: "The more I examine the universe and the details of its architecture, the more evidence I find that the universe in some sense must have known we were coming."[69]

God's Summary

Concise yet precise, the Genesis account summarizes each phase of creation with the simple statement, "And God saw that it was good." Indeed, how could these careful selections be better described? The last verse of Genesis 1 emphasizes the point, and God reminds us of an artist surveying His completed work: "God saw all that He had made, and it was *very good* (Genesis 1:31, emphasis added).

Carl: How about if we call the book, *The Cushy Cosmos?*

Fred: Not bad.

Carl: On second thought, most people aren't really satisfied with their lot in life and they don't feel they have it so cushy. They wouldn't think the book is for them.

Fred: Maybe most people need to understand more about what we've got going for us here.

Carl: The earth can get pretty rough sometimes.

Fred: Not compared to anywhere else we know about. Maybe this quote from NASA astronomer John O'Keefe will help to broaden our perspective:

> . . . to the astronomer, the earth is a very sheltered and protected place. There is a marvelous picture from Apollo 8 of the blue and cloud-wrapped earth, seen just at the horizon of the black cratered, torn and smashed lunar landscape. The contrast would not be lost on any creature; the thought 'God loves those people' cannot be resisted. Yet the moon is a friendly place compared to Venus, where, from skies forty kilometers high a rain of concentrated sulfuric acid falls toward a surface that is as hot as boiling lead. [70]

Apollo 11's view 10,000 miles away from the moon. Our nearest neighbor has none of the amenities of home. —*NASA*

Carl: Sounds like the place where people tell people they don't like to go.

Fred: O'Keefe went on to say that Venus is *friendly* compared to the cold and lonely vacuum that separates the stars, which is friendly compared to the crushing pressures of white dwarfs or the "unspeakable horror" of black holes or neutron stars.[71] He concluded:

> We are, by astronomical standards, a pampered, cosseted, cherished group of creatures. . . . If the Universe had not been made with the most exacting precision we could never have come into existence. It is my view that these circumstances indicate the Universe was created for man to live in.[72]

Carl: And your point is?

Fred: My point is that all that exacting precision that went into making this world is for *you.*

Carl: For me?

Fred: For each of us individual persons. Someone made a lot of special arrangements and took a lot of time so that each of us could be alive and experiencing this just-right world. All eternity past has been anticipating these few moments you're now experiencing. This is it. Don't blow it. Don't waste it.

Carl: Seize the day, huh?

Fred: That's right. There's tremendous purpose in your being here. By all rights, you should have been nothing for all eternity. Life is an incredibly rare gift.

Carl: I never looked at it just that way.

Shocking Summary Statements
and
Stimulating Conversation Starters

The non-theist admits to observing behavior that can be expressed in predictable, orderly laws of nature, not arbitrary laws of nature—which raises the question, "Why *are* there these orderly laws of nature? Why not chaos?" The Bible believer has a supernatural explanation. Does the non-theist have a natural one?

◆ If non-theists are at a loss to explain how something can come from nothing, then they are at a greater loss to explain how *nature* could come out of nothing, with all its highly specific, organized and purposeful accomplishments.

◆ When Fred Hoyle calculated the likelihood that carbon would have precisely the required resonance by *chance*, he said that his atheism was greatly shaken, adding: "A common sense interpretation of the facts suggests that a superintellect has monkeyed with physics."

◆ Carl Sagan admits: "It is easy to see that only a very restricted range of laws of nature are consistent with galaxies and stars, planets, life and intelligence."

◆ Stephen Hawking cites the ratio between the masses of the proton and the electron as one of many fundamental numbers in nature. He adds: "The remarkable fact is that the values of these numbers seem to have been very finely adjusted to make possible the development of life."

◆ Physicists speak of four fundamental forces in nature: gravity, electromagnetism, and the strong and weak nuclear forces. Physicist Richard Morris writes: "Every one of these forces must have just the right strength if there is to be any possibility of life."

◆ What are the chances that all the functional proteins necessary for life might form in one place by random events? Hoyle and Wickramasinghe calculated the odds at 1 in $10^{40,000}$, a number far removed from the realm of finite possibilities.

◆ Nobel laureate Francis Crick (who first cracked the DNA code) has concluded, like Hoyle and a number of others, that life could not have originated here on earth by any natural process. Experiment and calculation have led these physicists and biologists this far; but because of their refusal to consider the Biblical explanation, they have been forced to resort to the panspermia hypothesis: the idea that *life's genetic material was sent here from somewhere else.*

The second law of thermodynamics tells us that the universe is wearing down, becoming less ordered, so that less energy is available for work. But no law or theory tells us anything about how the universe got into its low entropy (highly ordered) state to begin with. Roger Penrose's calculations show that this highly ordered initial state for the universe is not something that could have occurred by chance.

Our universe is precariously balanced between expansion and collapse, between an expansion force that prohibits the universe from crashing back into itself early in the expansion, and enough gravitational force so that matter can collect into galaxies rather than disperse into gases. The ratio between the universe's actual density to the critical density is called Omega. George Smoot says, "The most minute deviation either side of an Omega of 1 consigns our potential universe to oblivion."

According to the findings of 20th-century physics, matter and antimatter must have been produced in equal amounts in any conversion of energy into mass associated with the creation event. But this forces physicists to acknowledge yet another finely tuned parameter to bring about our habitable universe. Particles of matter must have slightly outnumbered particles of antimatter by an extremely critical amount. Once again, it appears that a very precise selection was made.

Princeton physicist Freeman Dyson writes, "The more I examine the universe and the details of its architecture, the more evidence I find that the universe in some sense must have known we were coming." NASA astronomer John O'Keefe says, "It is my view that these circumstances indicate that the Universe was created for man to live in."

When you consider the many, purposeful choices that were made to bring about this universe, you realize that all eternity past has been anticipating these few moments you're now experiencing. There is tremendous purpose in your being here. Your life is an incredibly rare gift.

Footnotes for Chapter 9

1. Freeman Dyson, *Disturbing the Universe* (New York: Harper & Row, 1979), p. 250.

2. Howard J. Van Till, Davis A. Young, and Clarence Menninga, *Science Held Hostage—What's Wrong with Creation Science AND Evolutionism* (Downers Grove, IL: InterVarsity Press, 1988), pp. 21-22.

3. Colin A. Russell, *Cross-currents—Interactions Between Science and Faith* (Grand Rapids: William B. Eerdmans Publishing Company, 1985), pp. 60-61.

4. Ibid., p. 76.

5. Isaac Asimov, *In the Beginning* (New York: Crown, 1981), p. 11.

6. Albert Einstein, *Ideas and Opinions—The World as I See It* (New York: Bonanza Books, 1931), p. 40. Emphasis added.

7. Russell, p. 66.

8. Ibid.

9. L. Pearce Williams, "Faraday, Michael," *The Dictionary of Scientific Biography,* vol. 4, ed. by Charles Coulston Gillispie (New York: Charles Scribner's Sons, 1971), p. 531.

10. Russell, p. 76.

11. Ibid., pp. 66, 74.

12. Richard Morris, *The Fate of the Universe* (New York, Playboy Press, 1982), p. 155.

13. Steven Weinberg, "Life in the Universe," *Scientific American* (October 1994), p. 49.

14. Robert P. Kirshner, "The Earth's Elements," *Scientific American* (October 1994), p. 61.

15. Author interview with James Truran, May 6, 1994.

16. Fred Hoyle, *Engineering and Science* (November 1981). Cited by Owen Gingerich, "Let There Be Light: Modern Cosmogony and Biblical Creation," *The World Treasury of Physics, Astronomy, and Mathematics,* ed. by Timothy Ferris (Boston: Little, Brown and Company, 1991), p. 392.

17. Ibid., p. 393.

18. Dyson, p. 251.

19. Morris, pp. 154-155.

20. Ibid., p. 154.

21. Carl Sagan, *Cosmos* (New York: Random House, 1980), p. 260.

22. Eugene Linden, "How the Earth Maintains Life," *Time* (November 13, 1989), p. 114.

23. James Lovelock, quoted by Linden, p. 114.

24. Claude J. Allégre and Stephen H. Schneider, "The Evolution of the Earth," *Scientific American* (October 1994), p. 71.

25. Stephen W. Hawking, *A Brief History of Time—From the Big Bang to Black Holes* (New York: Bantam Books, 1988), p. 125.

26. Morris, p. 153.

27. Hugh Ross, *The Fingerprint of God,* second edition, (Orange, CA: Promise Publishing Co., 1989, 1991), pp. 121-122.

28. Morris, p. 153.

29. Dyson, p. 250.

30. Hawking, p. 125.

31. Ibid.

32. Fred Hoyle and Chandra Wickramasinghe, *Evolution from Space* (London: J.M. Dent and Sons, 1981), p. 24.

33. Fred Hoyle, *The Intelligent Universe,* (New York: Holt, Rinehart and Winston, 1983), p. 242.

34. Davies, *God and the New Physics* (New York: Simon and Schuster, 1983), p. 188.

35. Ibid.

36. Edward Kolb, quoted by Sharon Begley, "Science of the Sacred," *Newsweek* (November 28, 1994), p. 58.

37. Sharon Begley, "Science of the Sacred," *Newsweek* (November 28, 1994), p. 58.

38. Davies, p. 168.

39. Isaac Newton, cited by Davies, p. 164.

40. Davies, pp. 168, 178-179.

41. Morris, p. 152.

42. John D. Barrow and Joseph Silk, *The Left Hand of Creation—The Origin and Evolution of the Expanding Universe* (New York: Basic Books, Inc., Publishers, 1983), p. 206.

43. Morris, p. 152.

44. Steven L. Beyer, *The Star Guide—A Unique System for Identifying the Brightest Stars in the Night Sky* (Boston: Little, Brown and Company, 1986), dedication page.

45. Stephen Hawking, quoted by John Boslough, *Masters of Time—Cosmology at the End of Innocence* (New York: Addison-Wesley Publishing Company, 1992), p. 55.

46. George Smoot, interview with the author, May 6, 1994.

47. Stephen W. Hawking, *A Brief History of Time—From the Big Bang to Black Holes* (New York: Bantam Books, 1988), p. 122-123.

48. Dennis Overbye, *Lonely Hearts of the Cosmos—The Story of the Scientific Quest for the Secret of the Universe* (New York: Harper Collins, 1991), p. 356.

49. David Schramm, "Dark Matter and the Origin of Cosmic Structure," *Sky & Telescope* (October, 1994), p. 30.

50. Overbye, p. 356. See also Paul Davies, *Other Worlds* (New York: Simon and Schuster, 1980), p. 174.

51. George Smoot and Keay Davidson, *Wrinkles in Time* (New York: William Morrow and Company, 1993), p. 161.

52. Ibid., p. 163.

53. Smoot and Davidson, *Wrinkles in Time* (New York: William Morrow and Company, 1993), p. 190.

54. Jeremiah Ostriker, interview with author, July 13, 1994.

55. Richard Morris, *Cosmic Questions—Galactic Halos, Cold Dark Matter and the End of Time* (New York: John Wiley & Sons, 1993), pp. 121-123.

56. Dyson, pp. 250-251.

57. Ibid., p. 251.

58. Weinberg, p. 45.

59. Smoot and Davidson, p. 112.

60. Isaac Newton, *Philosophiae Naturalis Principia Mathematica* (1687), trans. by Andrew Motte (1729), Definition 5.

61. Ron Cowen, "Jupiter and Saturn: Rare in the Cosmos?", *Science News,* vol. 143, no. 13, (March 27 1993), p. 198.

62. Hawking, p. 174.

63. Ibid., pp. 140-141.

64. Ibid., p. 174.

65. Stephen Hawking, in letter to the author, July 12, 1994. I include the exact quote on p. 83.

66. Hawking, p. 174.

67. Heinz R. Pagels, *Perfect Symmetry: The Search for the Beginning of Time* (New York: Simon & Schuster, 1985), p. 243.

68. Hugh Ross, *The Creator and the Cosmos* (Colorado Springs, CO: NavPress, 1993), pp. 111-114.

69. Dyson, p. 250.

70. John A. O'Keefe, cited by Robert Jastrow, *God and the Astronomers,* second edition (New York & London: W.W. Norton & Company, 1992), p. 117.

71. Ibid.

72. Ibid., p. 118.

An edge-on spiral, galaxy NGC4565. —*Courtesy of Lick Observatory, University of California*

CHAPTER 10

Alternative Explanations
for Design

"To whom will you compare me?
Or who is my equal?" says the Holy One. . . .
"Present your case," says the LORD.
"Set forth your arguments," says Jacob's King.
—Isaiah 40:25, 41:21

A Skeptic's Questions:

I grant you that the universe has the *appearance* of being designed, but I've heard that there are actually some perfectly natural explanations for it. It's not as if the Creator in Genesis is the only explanation. Or are you trying to say that the evidence from design is so impressive that it compels belief in Him?

A Bible Believer's Response:

I would never say that the scientific evidence *forces* anyone to believe in God. So long as people have an imagination, they will always have alternatives to God as the Grand Designer. I would only ask that you honestly compare those nontheistic alternatives to the God of the Bible, and then judge for yourself which is more reasonable. I think you'll be surprised when you learn exactly what the best explanations are that naturalists have to offer. Which takes more faith to believe? That design points to some means by which humans created themselves? Or that design points to an infinite number of universes that somehow sprang out of nothing? Or that design points to a transcendent Creator?

The Anthropic Principle

How do scientists explain the precise selections of the last chapter? Many simply do not, recognizing either that science has no answers at present or that answers involving such explanations will always lie outside the realm of science. Others speculate on explanations that involve some form of the anthropic principle. Essentially, this principle claims that man himself has somehow caused the universe to take the necessary form to bring about his own existence. A milder way of stating this is that the universe is somehow put together in such a way that makes it just right for humans. After all, the constants of physics can actually be predicted (before observation) on this basis. Some have used the anthropic principle to say that the universe was created especially for humankind. Others use it to argue that humankind itself is somehow more directly responsible for its own creation.

The Weak Anthropic Principle (WAP)

(A Privileged Viewpoint)

Morris Aizenman, Executive Officer in the astronomy division of the National Science Foundation, expresses a view that is typical of many of the scientists I have interviewed. By taking a position that comes under the WAP category, he avoids both the embarrassment of having to explain design and the larger embarrassment of having to resort to any of the wild, science fiction-like explanations I am about to describe. The way to avoid these, apparently, is to give no explanation at all—except that these parameters *must* be set as they are, or we wouldn't be here. Referring to the critical cosmic constants, I asked him: "Do you feel a blind chance position *can* adequately explain these, or that these very 'lucky accidents' point to something more than luck?" Dr. Aizenman answered:

> My frank answer is that no one knows why the mass of the electron is what it is or the mass of any particle is the way it is or why the physical constants come about, why they happen to be. At a certain point, you say, "Why? Why? Why? Why is that just like that?" And the answer to that would be because if it wasn't like that, we wouldn't be here. Each person has a way of reacting to what they see as the extraordinary harmony of the way things are in this context. Some people attribute it to a Creator, to an intelligence, and I leave that to personal choice in that context. One might say that I lack a certain amount of wonder that it just happened to be just so, but I say, "If it weren't so, we wouldn't be here."[1]

To be more precise, the weak version of the anthropic principle explains these "unnatural selections" by saying that our human existence places us in a privileged time and place. In a universe that is sufficiently large, the right conditions for life might occur in certain times and certain, rare regions. Thus an intelligent observer should not be surprised if he finds himself in a time and place where the conditions are just right for his existence.

Most of our universe is *not* so hospitable to life. For many of the billions of years preceding our existence, conditions would not allow intelligent life, which requires the heavy element debris from first generation stars, etc. So our observation of a perfectly prepared cosmos is "explained" by our privileged, human point of view. The only times and places observers can exist to make these observations of a perfect universe will be when and where those perfect conditions exist.

But is this an explanation? As Alan Guth told me, "The anthropic principle is incredibly vague. You can use the anthropic principle, if you want, to explain almost anything. And it never gives precise predictions; it only explains after the fact that what you saw was, in some sense, acceptable."[2]

I don't dispute the truth of this weak anthropic principle—it's perfectly reasonable. But I must dispute its value in explaining any of the last chapter's "unnatural selections." The perfect conditions all need to be set up at the beginning of our universe—and we still have no natural explanation for such selections.

The weak anthropic principle underestimates how rare—in fact, how *impossible*—this privileged time and place must be, unless the allowed time or the number of regions is infinite. 20th-century cosmology teaches us that time is *not* infinite, but has a beginning. With regard to the concept that there may be many "regions" in our universe, each with its own physical laws, the concept abandons the WAP category and makes a not-so-subtle appeal to the multiple universe idea (SAP), discussed below. In any case, the idea that the physical constants may change from place to place rests solely on speculation, not observation.

Our universe was born with natural laws that do not appear to change. Thus most of our "unnatural selections" are not aided by adding more time. And for those that might be helped with more time, the time our universe offers—at most twenty billion years—is not nearly enough.[3] Clearly, nontheists need something stronger if they want a real explanation for the universe's fine-tuning.

The Strong Anthropic Principle (SAP)

(Multiple Universes)

Since our universe doesn't offer enough time for all these accidents to occur, some theorists have resorted to the thought that perhaps there are *many* universes, decreasing the odds against these accidents. In fact, if there are an *infinite* number of universes, then all of these accidents that produce our universe must have occurred in one of them. Such a notion is based on the idea that, given an infinite number of chances, *everything* will eventually happen. Some universes would even be very close to ours, except that in some of them Elvis Presley would have kicked his drug habit, gotten involved in Tennessee politics, and would now be serving as the President of the United States.

According to one version of SAP, every time there is more than one possible result for an action, the universe splits and forms a new universe for each result. One interpretation of quantum theory (discussed below) has prompted some physicists to speculate that elementary particles, which cannot have their momentum and position measured simultaneously, actually take an infinite number of positions, and each position exists in a different universe. According to this interpretation of an interpretation, each subatomic particle becomes real only when it is observed, and each particle may have a different observer in a different universe.

Thus universes diverge from one another like the branches of a spreading tree. Somewhere there must be a universe just like ours, except that last-minute negotiations between the U.S. and Japan in early December 1941 prevented the attack on Pearl Harbor. This one deviation from our history could begin a chain of events that would branch off farther and farther from ours: In that universe, the U.S. never entered World War II, with the result that Hitler's scientists developed the atomic bomb before the U.S., with the result that the Axis powers conquered the world and Nazi flags are now flying all over America.

This hypothesis, known as the "many worlds interpretation of quantum mechanics," was first proposed by an American physics student named Hugh Everett for his doctoral dissertation in 1957. It received further attention in 1973, when Stephen Hawking and Barry Collins at Cambridge published their findings that life could only exist in a universe where the galaxies were spewed out from the big bang at just the right rate to avoid re-collapse. Their calculations showed that this rate must be not only close, but exact, or galaxies could not exist. In fact, they found that the probability that this precise rate of expansion would be achieved was *zero*.

Hawking and Collins concluded that the only way out of this difficulty was to propose an *infinite* number of universes to balance the equation, so that *all* possible initial conditions would exist. Alternatively, Hawking speculated that there might be many separate regions in the same universe, each with its own physical constants.

Carl Sagan has written that perhaps an oscillating universe (a big bang followed by a big crunch followed by a big bang, etc.) would expand the possibilities, so that this habitable universe might eventually come about. But even in this case, he notes, the possibilities would not be infinite, since once a universe stumbles upon laws that allow it to continue its expansion, that would end all further possibilities.[4] And he adds, "If the laws of nature are unpredictably reassorted at the cusps, then it is only by the most extraordinary coincidence that the cosmic slot machine has this time come up with a universe consistent with us."[5]

Inflationary cosmologist Andrei Linde now proposes that his "chaotic" inflation scenario might result in many universes, each with its own physical laws. "According to this scenario," he says, "we find ourselves inside a four-dimensional domain with our kind of physical laws, not because domains with different dimensionality and with alternative properties are impossible or improbable but simply because our kind of life cannot exist in other domains."[6]

And Linde claims another possibility for producing multiple universes. Inspired by the ease with which he was able to create inflation models on a computer, he speculates on the possibility of creating an actual universe in a laboratory. The trick, he says, is to compress matter in such a way as to trigger inflation. Writing for the November 1994 *Scientific American,* Linde waxes godlike:

> And even if it is possible to "bake" new universes, what shall we do with them? Can we send any message to their inhabitants, who would perceive their microscopic universe to be as big as we see ours? Is it conceivable that our own universe was created by a physicist-hacker?[7]

Linde concedes that this process may not be possible; but if it is, then there might be a universe inside a universe inside a universe. Of course, many would say that Linde's ideas only work by adding together a long list of tenuous assumptions, and Linde himself admits that his "inflationary models are based on the theory of elementary particles, and this theory by itself is not completely established. . . . The inflationary theory itself changes as particle physics theory rapidly evolves."[8]

If other universes do exist (in finite numbers), what about the possibility that very different conditions might have simply resulted in very different types of life from the kind we know? Every physicist who has come to grips with the fine-tuning requirements for the formation of stars and the heavier elements has wondered about this possibility. Perhaps other worlds routinely come up with other forms of life based on other elements than the ones we use. In this case, life would not be such a rare thing, and the "just-right" conditions for life in this universe would not be so special. However, even Carl Sagan admits:

No other chemical element comes close to carbon in the variety and intricacy of the compounds it can form; liquid water provides a superb, stable medium in which organic molecules can dissolve and interact. . . . Certain atoms, such as silicon, might be able to take on some of the roles of carbon in an alternative life chemistry, but the variety of information-bearing molecules they provide seems comparatively sparse. Furthermore, the silicon equivalent of carbon dioxide (silicon dioxide, the major component of ordinary glass) is, on all planetary surfaces, a solid, not a gas. That distinction would certainly complicate the development of a silicon-based metabolism.[9]

"For the moment," concludes Sagan, "carbon- and water-based life-forms are the only kinds we know or can even imagine."[10]

Triffid Nebula, NGC 6514, in the constellation Sagittarius. Clouds of gas may be beautiful, but if that were all the universe had to offer, who would ever know? This naturally colorful cloud of hydrogen and helium (shown in its true colors at the top of our front cover) glows from the radiation of internal stars. Radiation pressure and stellar winds create a shock wave that pushes the gases outward. The dark lanes are opaque regions of thick gas and dust. —*National Optical Astronomy Observatories*

But what about the possibility of some sort of life composed of hydrogen and helium alone, or life without stars? Stephen Hawking considered the idea that universes without our special laws—and hence without stars or heavy elements—might produce some sort of life that even science fiction writers haven't thought of. When we begin moving beyond the dreams of science fiction writers, it may seem that we have moved too far; but surely some such drastically different form of life is theoretically *possible*. Hawking writes:

> Nevertheless, it seems clear that there are relatively few ranges of values for the numbers that would allow the development of any form of intelligent life. Most sets of values would give rise to universes that, although they might be very beautiful, would contain no one able to wonder at that beauty. One can take this either as evidence of a divine purpose in creation and the choice of the laws of science or as support for the strong anthropic principle."[11]

These then, appear to be our choices, given the facts of physics we observe: multiple universes or divine design. Physicist/mathematician Paul Davies notes that "one may find it easier to believe in an infinite array of universes than in an infinite Deity, but such a belief must rest on faith rather than observation."[12]

The fact that this multiple universe idea was for a time taken seriously by a number of scientists is good evidence for the lack of reasonable alternatives to God as the Grand Designer. Today, the multiple universe hypothesis is unacceptable to most physicists. Even John Wheeler, its greatest promoter in the 50s, 60s, and 70s doesn't believe in it anymore. Yet it continues to be the main explanation physicists fall back on when I ask them to explain how, against incredible odds, all the conditions are perfectly balanced to create life-sustaining stars and galaxies in this universe. They may not believe it, but it makes for the best naturalistic answer they have.

Theoretical physicist Jeremiah Ostriker used the multiple universe argument in order to explain to me how the universe might have achieved the critical density to last as long as it has, against incredible odds. "There are many inflationary models in which there are separate bubbles," he said, "and you'll find yourself in one or another one. And you're more likely to find yourself in a bubble which lasts a long time."[13] But Hawking found that there would have to be, not just many, but an *infinite* number of universes to come up with a cosmos that meets this one condition. And Dr. Ostriker only used this argument to try to satisfy the critical density condition. He didn't address the other "numerical coincidences" that I asked him about.

Carl: So what you're saying is that, according to the Copenhagen interpretation of Popular Mechanics—

Fred: That's *quantum* mechanics.

Carl: Yeah, according to that, *everything* happens in some universe?

Fred: Some philosophers say that maybe each time you decide where to go out to dinner, the universe splits into two new universes and in one universe your wife gets her way and in the other universe you get your way. So it really doesn't matter if you get your way or not, because somewhere, you're getting your way all the time.

Carl: Hey, you know this could take the pressure off a lot of marriages. Now you're getting relevant.

Fred: Well, it's more philosophy than science.

Carl: I wonder if in some universes my wife runs out of honey-do's.

Fred: The point is, the fact that physicists seriously resort to multiple universes shows how desperate—

Carl: Wait! I've got a new book concept! *The Quantum Stress Reducer.* I could write this. "Wives, in some universes you *succeed* in making your husband into everything you think he ought to be. So don't worry if you don't quite get him there in this one. There'll be other chances." Hey, I could sell this as the perfect gift for husbands to get their wives. How's that for a hook?

Notice that most of the hottest issues that cosmologists grapple with today involve questions having to do with finely-tuned conditions (questions involving missing mass, dark matter, the density of the universe, element production, fluctuations in the microwave background, etc.). Flashier issues, like the question of extraterrestrial life, may continue to receive more attention in the general media. But the actual scanning of stars for signs of intelligence is a relatively unimaginative exercise that does much less to promote knowledge of our universe. As physicist Richard Morris states, "The really intriguing question may very well be not, 'Is there other intelligent life in the universe?' but rather, 'Why is the universe so hospitable to life in the first place?'"[14]

The Bible's answer is exactly what scientists should have come to expect: that God exercised tremendous wisdom, power, and care in forming such a perfect universe for us. Jeremiah said, "God made the earth by His power; He founded the world by His wisdom and stretched out the heavens by His understanding" (Jeremiah 10:12). Isaiah said that He "did not create it to be empty, but formed it to be inhabited" (Isaiah 45:18). Paul wrote that God's "purpose" for us was planned even "before the beginning of time" (2 Timothy 1:9).

Natural laws that are so precisely contrived to sustain us show His long-sighted faithfulness toward us, even from ages past (Psalm 119:90-91). The superior intelligence that physicists recognize behind nature and its laws is called "wisdom" in the Bible. Proverbs 8 personifies wisdom as the craftsman at God's side throughout the creation process, ending with His artisan's delight in finally creating humans:

> I was appointed from eternity, from the beginning, before the world began . . . . before He made the earth or its fields or any of the dust of the world. . . . Then I was the craftsman at His side. I was filled with delight day after day, rejoicing always in His presence, rejoicing in His whole world and delighting in mankind.

(Proverbs 8:23,26,30-31)

The Participatory Anthropic Principle (PAP)

(We Created Ourselves by Observing Ourselves)

This next form of the blind chance argument makes use of what appears to be the *ultimate* game of chance: quantum mechanics. Some people have tried to use quantum mechanics to say, as astronomer George Greenstein put it, "the universe brought forth life in order to exist . . . the very cosmos does not exist unless observed."[15] This notion is founded on what is known as the Copenhagen interpretation of quantum mechanics, which says that there is no reality in the world until one observes it.

The Copenhagen interpretation is one of a number of philosophical theories to explain why a quantum particle can never have both its position and its momentum measured at the same time. This curious fact is called the Heisenberg uncertainty principle. The German physicist, Werner Heisenberg, received a Nobel prize in 1932 for his development of quantum mechanics, a significant deviation from Newtonian physics, like Einstein's principle of relativity. Relativity becomes useful when exploring phenomena at velocities that approach the speed of light, which defy classical explanation. Similarly, quantum mechanics becomes useful when studying extremely small-scale phenomena, where the dimensions are small compared to even the smallest wavelengths of light. Once again, activities in this special realm cannot be explained by ordinary mechanics.

In 1927, Heisenberg discovered that measuring a quantum particle's position makes it impossible to measure its velocity, and measuring its velocity makes it impossible to narrow down its position to an actual point from

Werner Heisenberg lectures at OSU in 1954. —*Courtesy AIP Emilio Segré Visual Archives, Lande Collection*

a range of possible places where it might be. I have been told that hotels all over Germany display signs saying, "Heisenberg might have slept here."

This range of positions where the particle might be found is called its "possibility distribution" or "wave function." Quantum theory can help us make predictions about quantum particles based on statistical laws of probability. But as long as an observer is dealing with single particles at the subatomic level, he is limited to choosing between learning about the particle's position or its energy level—never both.

Some have taken this to mean that the very act of observation is what gives the particle reality. After all, the particle does not appear to be in any particular place until we observe it. Perhaps, say some philosophers, cause and effect break down at this small-scale level. Those who draw these conclusions are following in the footsteps of Danish physicist Niels Bohr, who may have been influenced by his own Hindu beliefs when he proposed his famous "Copenhagen interpretation."

Actually, the Heisenberg uncertainty principle does not imply that the particle is causeless, but merely that science has run up against a limit in its ability to predict an effect. An intelligent observer is not necessary to give the particle reality; a photographic plate will detect the particle just as well.

The idea that mere observation *causes* anything not only stretches logic, but it violates Einstein's speed limit, the rule that nothing in the universe can travel faster than light. If the Copenhagen interpretation were true, it would mean that the mere act of looking through a telescope could alter events happening many billions of years in the past. This thought has given some Copenhagen enthusiasts the chance to speculate that perhaps man's very act of observation has caused his own creation, along with all the conditions necessary for human life.

Of course, such an idea just creates a "which came first, the chicken or the egg" problem that goes on forever; it does not get around the need for a first cause. Also, the quantum uncertainties are true only of micro systems, not of the larger systems, which are made up of many multiples of quantum particles. As we progress to larger systems from the micro world, the uncertainty of a system quickly reaches zero. Physicists have never observed any quantum effects in the visible world. Quantum mechanics has no bearing on the operation of a car, for example. The Copenhagen interpretation simply cannot be applied to the whole universe in this way. Very few scientists take seriously the notion that the universe could be so perfectly designed merely by man's observation of it.

Yet, in attempting to apply quantum mechanics to the whole universe, many of today's physicists continue to search for some way to explain the universe's behavior—and perhaps even its existence—in terms of a quantum mechanical wave function.[16] Some speculate that if the behavior of an electron can be described as a wave of possibilities (a wave function) that cannot be nailed down until an observer comes along, perhaps the initial conditions of the whole universe may be treated as a wave function. Perhaps the whole universe had no particular behavior—or according to the Copenhagen interpretation, it had no *existence*—until it was observed. But who would be outside the universe to observe it? Does this not begin to sound something like a transcendent God? The many physicists who have attempted to apply quantum mechanics to the entire universe are thus driven back to the supernatural explanation they were striving to avoid.

The Final Anthropic Principle (FAP)

(Our Descendants Will Become God)

In their 1986 classic, *The Anthropic Cosmological Principle*,[17] astronomer John Barrow and mathematical physicist Frank Tipler tried to calculate how long it might take for the critical conditions we observe in the universe to come together by nontheistic explanations. After examining ten conditions for life, they found that all the time in the world did not give them what they needed to explain the development of human beings by natural means. Even when they gave nature all the lucky breaks they could, their calculations showed that by the time any *one* of these ten precise conditions might be met by natural means, the sun would cease to be a main sequence star, turning into a red giant and burning up the earth. In the end, they concluded that there must be an intelligence to make the intelligent selections for human life.

Driven to this position, but unwilling to accept a more obvious biblical explanation, Barrow and Tipler speculated on ways (like WAP, SAP, and PAP) that humankind itself might *be* that intelligence. Their own preferred hypothesis was that life may evolve to such an advanced degree that it will become an all-knowing, all-powerful, omnipresent god. Having amassed such powers, this evolved god may then be able to create in the past. They called this concept the Final Anthropic Principle (FAP).

In *The New York Review of Books*, Martin Gardner considered the dubious nature of all these anthropic principles and concluded that he could come up with a more appropriate acronym for the last one: "What should one make of this quartet of WAP, SAP, PAP, and FAP? In my not so humble opinion I think the last principle is best called CRAP, the Completely Ridiculous Anthropic Principle."[18]

In Frank Tipler's 1994 book, *The Physics of Immortality*,[19] he develops his FAP further, attempting to make theology into a branch of physics. Though he believes in a closed universe (and his entire argument hinges upon this), he says that life must exist until the end of time. But in order for life to survive in the hot, dense time when the universe will contract once again to a point, our descendants must evolve into very different, very advanced beings. Before the end, these super-computer-like beings will achieve the ability to process an infinite amount of information. At that point, which Tipler calls the "Omega Point," they will assume the role of God. By processing an infinite amount of thoughts at infinite speed, this evolved Supreme Being will redefine time, ensuring an eternity for itself and for every being that ever existed before. Viewing the human soul as a program that can be replicated, Tipler proposes that this infinite intelligence will emulate (or resurrect) each of us, so that we might be appropriately rewarded or reformed.

The book's promoters made much of the fact that Frank Tipler is "a former atheist," as if his mathematical musings on God and resurrection has turned him into a man of faith. Actually, of course, Tipler has studiously avoided a transcendent God—the God of the Bible. By avoiding the Creator from *outside* the universe, he has avoided the one solution that could explain it.

Once again, the fact that physicists are willing to consider such outrageous explanations only illustrates how impossible it is for them to avoid evidence of design. Evidence of design points to intelligence. Each individual must decide whether it is more reasonable to credit this intelligence to God or to ourselves.

The Search for Extraterrestrial Intelligence

The thought that humans might be the only form of intelligent life in the universe strikes most scientists as anti-Copernican and downright unscientific. If life is not an inevitable result of the right conditions, humans are elevated to a suspiciously special place. In order to conform to the usually-held assumption about biological evolution, many scientists have proposed that the universe must be "teeming with extraterrestrials."[20] Mount Wilson astronomer Robert Jastrow told me that if this is true, then we should expect to receive signals from extraterrestrial life in the near future: "If life is common, we'll be hearing from those guys soon, because we're a very conspicuous part of the universe right now. Our TV and radio is spread out all around us."[21] And Robert Arnold of the SETI Institute wrote me:

> These electromagnetic artifacts of daily commerce, entertainment and defense give the earth a distinct radio frequency signature that is brighter than the Sun. Distant technological civilizations may have detected these transmitters. It is reasonable to assume that advanced technological civilizations will be aware of radio physics, and there is only one radio spectrum. It is the same everywhere in the universe.[22]

But to most scientists who have examined the subject in depth, the expectations raise an embarrassing question: "Why haven't we picked up their signals yet?" Voicing the opinion of many scientists, Italian physicist Enrico Fermi (who directed the design of the first atomic reactor) said that by this point in our universe's history, there should be many thinking beings throughout the universe. Given the fact that, once intelligence evolves, it takes such a short amount of time to develop sophisticated technology (apparently just a few thousand years in our case), there should be many civilizations far more advanced than ours. Certainly they would explore and colonize other planets of the galaxy, and ours would offer an ideal place with all the right conditions for life. And so Fermi asked the great question that shouldn't have to be asked, if all this were true: "So where *are* they?"[23]

The fact that NASA spent sixty million dollars on SETI indicates the seriousness of this issue to scientists. To rephrase Fermi's question: At least, why aren't we picking up their radio signals? Could it be that science has made some wrong assumptions somewhere along the line?

Nobel laureate Francis Crick, known for his co-discovery of DNA and for cracking the genetic code, writes:

> An honest man, armed with all the knowledge available to us now, could only state that in some sense, the origin of life appears at the moment to be almost a miracle, so many are the conditions which would have had to have been satisfied to get it going.[24]

In his book, *Are We Alone? The Possibility of Extraterrestrial Civilizations,* the prominent physicist James Trefil concludes:

> For, as we have seen in this book, the evidence we have at present clearly favors the conclusion that we are alone. From the formation of the sun as a single G star to the evolution of the earth's atmosphere to the conditions of the earth's recent climate, everything points to the same conclusion—we are special.

> But we are living on an insignificant speck of rock going around an undistinguished star in a low-rent section of the galaxy. We are not the center of the universe.

> Maybe so, but we are special.

> . . . . If I were a religious man, I would say that everything we have learned about life in the past twenty years shows that we are unique, and therefore special in God's sight. Instead I shall say that what we have learned shows that it matters a great deal what happens to us.[25]

Why We're Special—
With or Without a Recent Creation

To some, the evidence creates a paradox—or even a contradiction. How could there be so much evidence for our being special and at the same time so much evidence that we are insignificant? Doesn't science teach us that we live in a time and place that are mere specks in a universe that spans billions of light-years and billions of years in age?

The recent creation position gives us one way to be special. Because of its rigidity, the way involves limiting the means of God's creativity. It assumes that, in order for creation to be special (i.e., supernatural or miraculous), God cannot use time and He cannot use natural processes. Rather, this view ensures the special nature of God's creative acts by holding that

God created everything during a 144-hour period, about ten thousand years ago. So much for having to worry about the billions of "wasted" years that preceded us.

Those who take the discoveries of science more seriously hold that God *did* use vast periods of time and several generations of stars to create the present order. But doesn't this create problems for those who believe that God made the universe especially to be inhabited by *us*? After all, the honest, unassuming response of anyone who bothers to learn about our universe is humility—we're amazed at its enormity. All our concerns are dwarfed when we gaze up at the night sky. When David did so, he expressed the opposite of the anthropic principle: "When I consider Your heavens, the work of Your fingers, the moon and the stars, which You have set in place, what is man that You are mindful of him, the son of man that You care for him?" (Psalm 8:3-4).

Such a statement is more in keeping with what has become the perspective of modern science. Science has gradually progressed from a geocentric view to a true appreciation, only in this century, of the seeming insignificance of even our galaxy among billions of galaxies. Those who believe in these more imaginative forms of the anthropic principle are running against the grand progression of scientific thought. To view mankind as the natural center is to revert to a view that science left behind long ago.

But for some, this raises questions that seem to create problems for the biblical view: If the universe is so vast, why *should* we feel special? If the universe was designed with human beings in mind, why doesn't it just have one star (our sun) and one planet (our earth), or at most just one galaxy (our Milky Way)? Why waste all those other billions of stars and galaxies? Just to look pretty? Why can our telescopes see billions of light-years into space? Why should God bother making the universe so *big*? And why is there so much "wasted" *time* before us?

Whether or not they like the idea of divine design, astronomers today can answer all of these questions in a way that assures us of the Creator's particular care and preparation for us right from the beginning. Speaking of the enormity of our universe, astronomers John Barrow and Joseph Silk explain, "No astronomer could exist in one that was significantly smaller."[26] They point out that the elements necessary for life—carbon, nitrogen, oxygen, phosphorus, and silicon—were not formed and available for use until after the first generation of stars had slowly cooked the simple elements of the big bang into the heavier elements. No life can come from hydrogen and helium alone—hence the need for a completed first generation of stars and an ancient universe.

God, Time's Relativity, and Quantum Mechanics

Relativity theory tells us that time is not the same to everyone. The Bible tells us that time is not the same to God as it is to us. "With the Lord a day is like a thousand years, and a thousand years are like a day" (2 Peter 3:8). To the Lord, an age might be like a mere watch in the night (Psalm 90:4). Should vast stretches of time impress Him the way they impress us? The biblical God is outside of time, so what difference does it make to Him if He takes a day to create something, or a thousand years, or a billion years? Time only matters to God when it matters to us mortals, because He knows our time is limited. But before He created us, why shouldn't He take billions of years, if He wanted to, to set up the universe?

Is there any place in the Bible that says He created the universe instantaneously? Is our universe any less of a miracle if God chose to create it over vast stretches of time, which, when measured by our present earthly rotation periods around the sun, seems to dwarf our lifetimes into insignificance?

How ironic it is that one of the implications of quantum mechanics today is that our universe must indeed be the result of a supernatural miracle—*not because it was created in an instant, but because it has continued for billions of years.* According to quantum mechanics, once space is created, a quantum fluctuation can produce particles from space (instantaneous creation), but these particles must disappear within an incredibly small fraction of a second, called Planck time (10^{-43} second). Every fraction of a second that transpires beyond that first instant of creation adds to the miracle of a universe that, according to quantum physics, never should have lasted beyond Planck time (see Robert Gange's discussion of this fact on page 318 in Bonus Section #2). *Instantaneous creation is thus less a miracle than a creation that took place over billions of years.*

A considerable amount of time is involved in the slow process of cooking the elements needed for life in the interior of a star, and after a star dies and explodes, more time is needed for the incorporation of these elements into a planet like earth. And given the fact that the universe maintains its stability through its expansion, more time results in a bigger universe. Barrow and Silk conclude:

Hence, we realize that for there to be time to construct the constituents of living beings, the universe must be more than a billion years old and consequently more than a billion light-years in size. . . . The universe would have to be just as large as it is to support even one lonely outpost of life.[27]

Looking up at the heavens *is* supposed to humble us, perhaps to remind us that we're not quite the master and measure of all things we would like to be, and certainly to teach us that whoever created such a spectacle is worthy of all the praise we can give. Psalm 66:5 proclaims, "How awesome [are] His works in man's behalf!"

Creation should inspire not only humility, but gratitude toward a Creator who would go to such amazing lengths in order to prepare a place for us. Indeed, if He is preparing a place for our future (John 14:2), we might reasonably infer that He did some preparing before He gave us *this* place. The more we learn of His universe, the more we understand why Psalm 19 says that the heavens are declaring, proclaiming, pouring forth speech—communicating His glory and His knowledge with a voice that goes out into all the earth.

Natural revelation tells us not only how small we are, but how great He is, whether we glance casually at the sky or devote our lives to discovering the incredible, unnatural selections that have brought about such order. Such careful selection tells us of His care for us. Such unfathomable physical mysteries tell us of His wisdom. Such mathematical precision tells of His perfection. Such immense energy speaks of His power. Such symmetry and elegance speak of His love of beauty.

These attributes—aesthetic sense, well-directed power, perfection, tremendous wisdom, and care—point to the personhood of a transcendent, biblical God, not to the unplanned, mindless accidents of blind chance.

Shocking Summary Statements
and
Stimulating Conversation Starters

 Which takes more faith to believe? That design points to some means by which humans created themselves? Or that design points to an infinite number of universes that somehow sprang out of nothing? Or that design points to a transcendent Creator? These are our alternatives.

The weak version of the anthropic principle tries to explain the universe's fine-tuning by saying that our human existence places us in a privileged time and place. But this "explanation" explains nothing about how all the perfect conditions were set up at the beginning of our universe. Without these carefully selected conditions, no time or place could ever produce or sustain intelligent life.

The fact that naturalists must resort to the multiple universe idea shows how impossible it is for them to explain the design of our universe by any natural means. If the number of universes is infinite, then *everything* happens, including this wonderful universe that so perfectly mimics divine purpose. Of course, this argument can be used to defend *anything*, including the idea that Elvis Presley is alive and now serving as President of the United States.

Physicist Paul Davies notes that "one may find it easier to believe in an infinite array of universes than in an infinite Deity, but such a belief must rest on faith rather than observation."

If, as George Greenstein put it, "the very cosmos does not exist unless observed," then one must wonder who is outside of the cosmos to observe it. Such a Creator from outside the universe begins to sound like a transcendent God. Those who would try to explain the cosmos as a quantum wave function are thus driven back to the supernatural explanation they were trying to avoid.

Concluding that there must be an intelligence to make the intelligent selections for human life, Frank Tipler has written that our descendants will evolve into God, who will then create us, working in the past. Once again, the idea ignores the logical need for a cause to *precede* the effect.

Scientists from Einstein to Hoyle have acknowledged that the evidence of design points to intelligence. Each individual must decide whether it is more reasonable to credit this intelligence to God, to the universe, or to ourselves.

Believing that life is an inevitable result of the right conditions, many scientists believe that the universe must be teeming with extraterrestrial intelligence. There is just one radio spectrum that they could use. So the question naturally arises: Why, with all our efforts, can't we pick up any of their communications signals? Could it be that our assumptions about life's inevitability are wrong?

 Every fraction of a second that transpires beyond the first instant of creation adds to the miracle of a universe that, according to quantum physics, never should have lasted beyond Planck time. The idea of instantaneous creation is thus less of a miracle than a creation that took place over billions of years.

 Why would God make us such a small part of such a huge and ancient universe? Astrophysicists are perfectly aware of the answer. Carbon, nitrogen, oxygen, etc. were not formed until after the first generation of stars had slowly cooked the simple elements into these heavier elements necessary for life. Billions of years of expansion results in a universe that is billions of light-years in size. Astronomers Barrow and Silk conclude, "The universe would have to be just as large as it is to support even one lonely outpost of life."

 The immensity of creation should inspire not only humility, but gratitude toward a Creator who would go to such amazing lengths in order to prepare a place for us. Natural revelation tells us not only how small we are, but how great He is: how great His power, His wisdom, His perfection, His beauty, His love.

Notes for Chapter 10

1. Morris Aizenman, interview with the author, May 3, 1994.

2. Alan Guth, interview with the author, June 8, 1994.

3. David Foster, *The Philosophical Scientists* (New York: Dorset Press, 1991), p. 83.

4. Carl Sagan, *Cosmos* (New York: Random House, 1980), p. 260.

5. Ibid.

6. Andrei Linde, "The Self-Reproducing Inflationary Universe," *Scientific American* (November 1994), p. 55.

7. Ibid., p. 53.

8. Ibid., pp. 53-54.

9. Carl Sagan, "The Search for Extraterrestrial Life," *Scientific American* (October 1994), p. 93.

10. Ibid.

11. Stephen W. Hawking, *A Brief History of Time* (New York: Bantam Books, 1988), p. 125.

12. Paul Davies, *God and the New Physics* (New York: Simon and Schuster, 1983), p. 174.

13. Jeremiah Ostriker, interview with the author, July 13, 1994.

14. Richard Morris, *The Fate of the Universe* (New York: Playboy Press, 1982), p. 150.

15. George Greenstein, *The Symbiotic Universe: Life and Mind in the Cosmos* (New York: William Morrow, 1988), p. 223.

16. Steven Weinberg, "Life in the Universe," *Scientific American* (October 1994), pp. 48-49.

17. John D. Barrow and Frank J. Tipler, *The Anthropic Cosmological Principle* (New York: Oxford University Press, 1986).

18. Martin Gardner, "WAP, SAP, PAP, and FAP," *The New York Review of Books* (May 8, 1986), pp. 22-25.

19. Frank J. Tipler, *The Physics of Immortality—Modern Cosmology, God and the Resurrection of the Dead* (New York: Doubleday, 1994).

20. Robert T. Rood and James S. Trefil, *Are We Alone? The Possibility of Extraterrestrial Civilizations* (New York: Charles Scribner's Sons, 1981), p. 247.

21. Robert Jastrow, interview with the author, December 15, 1994.

22. Robert Arnold, SETI Institute, letter to the author, August 2, 1994.

23. Enrico Fermi, cited by Francis Crick, *Life Itself* (New York: Simon and Schuster, 1981), p. 14.

24. Francis Crick, *Life Itself* (New York: Simon and Schuster, 1981), p. 88.

25. Rood and Trefil, pp. 251-252.

26. John D. Barrow and Joseph Silk, *The Left Hand of Creation—The Origin and Evolution of the Expanding Universe* (New York: Basic Books, 1983), p. 205.

27. Ibid.

Great Nebula in Orion, M42. We see it as a "star" in the sword of the constellation Orion. This spectacular gas cloud has been excited to luminosity by the ultraviolet radiation from several extremely hot stars in a cluster buried deep within the nebula. The nebula covers 120 light-years and is 1,500 light-years away.
—*National Optical Astronomy Observatories*

CHAPTER 11

Implications of Design

Shout for joy, O heavens; rejoice, O earth;
  burst into song, O mountains!
For the LORD comforts his people and will
  have compassion on his afflicted ones. . . .
"Can a mother forget the baby at her breast
  and have no compassion on the child she has borne?
Though she may forget, I will not forget you!
See, I have engraved you on the palms of my hands. . . ."

(Isaiah 49:13,15,16)

A Skeptic's Questions:

I suppose all this evidence for design you've been citing has been leading up to some great conclusion about the meaning of life. But can't a person take the naturalistic position and create his own meaning in life?

And I suppose you're going to say that design implies a Designer, and we've all sinned against His absolute moral standards. But actually, don't we just get our moral values from society? Aren't moral standards relative, according to time and place?

A Bible Believer's Response:

If moral values are relative, according to time and place, then who's to say there should be any moral values at all? Who's to say that a society that condones rape and torture (as some have) isn't right? If you believe that rape and torture are always wrong, how do you know that?

The most important thing we should note about the evidence for design is that it brings good news. To understand why, consider first the implications of an accidental universe. If we humans are mere accidents in a purposeless cosmic scheme—if we live and die only to have all memory of us wiped out within a few generations—what person can honestly say he's comforted by or even fully satisfied with that? Obviously, no matter how hard we try to *be* somebody or to create meaning for ourselves, our lives cannot have *lasting* meaning or *lasting* value in an accidental universe.

The Values and Meaning Offered by the "Chance" Position

Those who look at the universe as a chance happening have difficulty in explaining how life has purpose or value. Yet they try. Carl Sagan presents himself as a champion of the blind chance, materialist position: "I am a collection of water, calcium and organic molecules called Carl Sagan. You are a collection of almost identical molecules with a different collective label. But is that all? Is there nothing in here but molecules?"[1]

Sagan would like his reader to believe that the answer is yes, but that this should not deter us from feeling "elevated." He tries to evoke a *feeling* of meaning without any facts to suggest its objective reality: "Some people find this idea somehow demeaning to human dignity. For myself, I find it *elevating* that our universe permits the evolution of molecular machines as intricate and subtle as we."[2]

In an apparent effort to create meaning or value, Sagan speaks of the "loyalties"[3] we owe to our species and our planet, and of the "obligation to survive"[4] which we owe to the Cosmos that created us. We can't argue with his good intentions in exhorting us to take better care of our planet, to take our "stewardship"[5] seriously. But we must wonder, in a Sagan universe, whose stewards *are* we?

If there is no Designer, then the symmetry and beauty we observe serve no end except to make the universe even more pathetic than it would be without such harmonious laws. Physicist Steven Weinberg, another renowned spokesman for the blind chance position, writes, "The more the universe seems comprehensible, the more it also seems pointless."[6]

What kind of value could people have in a pointless universe? Webster's dictionary defines "value" as "relative worth or importance." When you think about it, all valuable things have their value only because of their relation to something else, outside of themselves. Diamonds are just pieces of compressed carbon and would be worthless, no matter how rare, if they remained by themselves and never had a relationship with something outside of themselves. Gold, silver, gem stones—none of them has any value in themselves—until we bring people into the picture. Gold needs something other than gold to give it value.

If our universe came about by some strange fluke and there is nothing outside of it, no purposeful Creator beyond its time and space to value it or give it meaning, then it must *remain* without meaning. The universe can't generate its own meaning or value any more than a rare rock sitting on an uninhabited planet can ever be valuable sitting there all by itself.

Our very lives cannot be valued apart from relationships. Some find their value in their relationship to things—money, houses, cars. Some find their value in their relationships to people—family, friends, audiences. But what if we outlive our money or our friends—does that mean our lives have lost all value? The problem with people or things is that they don't *last*. Only something that lasts can give our lives *lasting* value.

In the same way, *moral* values cannot spring from nothing. Materialists are not being consistent when they claim they can have moral values without God. Anyone who has ever raised a child knows that it is reasonable to wish to instill in that child some sort of moral values. Most parents are not content to watch their children grow into self-centered, drug abusing, violent, sex offenders. Most are not even content with lazy slobs. But is it logical to ask the child to follow standards that come from nowhere? At some point the child is bound to ask a very good question: "Why?"

Of course, most nontheists will argue that they certainly *can* have moral values apart from the theist's standard of reference. For those who take the blind chance position and still wish to hold to some system of ethics, they must admit that right and wrong are determined, if not by God, then by *people,* who may differ over what is right and what is wrong. One person's right might be another person's wrong. To slaughter six million Jews may be terribly wrong to most but right and consistent with the world view of Adolf Hitler. Who is to say what's actually wrong?

Here the blind chance position begins its run down a blind alley, for when its proponents try to claim that there is a better standard than individuals—and yet they must avoid anything too much like an objective, God-like standard—they will usually settle on a sensible intermediate, like society. But society has at times upheld racism, slavery, even human sacrifice. If the blind chance proponent says that society was obviously wrong in those particular instances, we must ask, *Wrong* relative to what? By using that word, the materialist acknowledges that there must be a higher standard by which to judge.

On a practical level, T. de Witt Talmage challenged blind chance advocates over a century ago: "Try your scientific comfort on those parents who have lost their only child. You come in, and you talk to those parents about 'selection,' and about the 'survival of the fittest'. . . . Try that consolation . . . . The American people are finding out that worldly philosophy and human science as a consolation in time of bereavement are an illimitable, outrageous, unmitigated, and appalling humbug."[7]

Much more recently, agnostic Steven Weinberg simply wrote: "I do not for a minute think that science will ever provide the consolations that have been offered by religion in facing death."[8]

Design Drives Scientists to a Designer

Astronomers Barrow and Silk conclude that "A modern perspective on cosmic evolution leaves little to chance."[9] So what alternatives does this leave modern scientists? Physicist Barry Parker ends one of his books with the admission that the creation event we think we know so much about is still impossible without a Creator to explain the ready-made laws of nature: "Who created these laws? There is no question but that a God will always be needed."[10]

Famed physicist and mathematician Sir James Jeans, whose formulas led to our modern theories of galaxy formation, summed up his discoveries this way:

> The universe begins to look more like a great thought than like a great machine. Mind no longer appears as an accidental intruder into the realm of matter; we are beginning to suspect that we ought rather to hail it as the creator and governor of the realm of matter. . . . We discover that the universe shows evidence of a designing or controlling power that has something in common with our own individual minds . . . .[11]

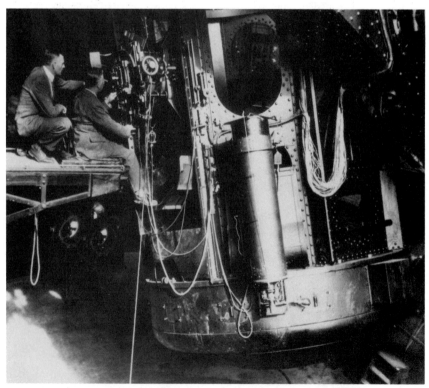

Edwin Hubble (left) and James Jeans sit on the Cassegrain platform next to Mount Wilson's 100-inch telescope in 1931. —*Courtesy of The Huntington Library.*

Astrophysicist Arthur Eddington came to a similar conclusion: "The idea of a universal mind or Logos would be, I think, a fairly plausible inference from the present state of scientific theory."[12]

In his book, *The Philosophical Scientists,* David Foster says, "I have found few, from Einstein to Schrödinger, who at some stage or another did not have to introduce God."[13] Foster speaks of the progressive demise since 1900 of scientific conclusions (particularly Darwinism) which had led scientists to speculate on the importance of "chance" in the order we observe. Chance or necessity cannot account for recent findings in microbiology, such as DNA coding or the specificity of hemoglobin. As in physics, modern biochemistry no longer allows for purely naturalistic explanations. Foster says that the old "chance" position is now "replaced by a new idea of the importance and dominance of *specificity* in the universe, with the inevitable implication that GOD EXISTS."[14]

Which God?—The Scientist's Choice

But when we turn to the question, "What *kind* of a God best explains the facts?" we begin to wonder if all this scientific progress isn't still at the mercy of the fundamental *goofiness* of human nature. By this (one of the few theological terms I have decided to include in this book) I mean only to point out that all humans are at times subject to a debilitating penchant— shared equally by scientists and laymen—to become so hard in the heart that they become soft in the head.

Cosmologists must make the same choices as the rest of humanity. In the matter of deciding who's running the universe, we all have three basic options: (1) the universe itself, (2) mankind, or (3) God. A person's choice on this matter will have obvious repercussions on the matter of who will run his or her life. Knowing this, isn't it possible that scientists, like all members of our race, may be predisposed to make their choice on the basis of factors other than pure logic?

Those unwilling to accept the implications of giving God control of their lives are free to choose either of the other two options, but not without giving up a bit of logic and taking on a great deal of blind faith. Remember, science still has no theory to explain how either the universe or humanity could (1) create itself or (2) design itself.

Romans 1:22,25 describes, in unflattering terms, those who make these choices in spite of this little logic problem: "Although they claimed to be wise, they became fools . . . . They exchanged the truth of God for a lie, and worshipped and served created things rather than the Creator." In suggesting that this passage might apply to certain scientists, are we being too hard on them? After all, modern scientists do not actually make the sun and the stars their gods, do they?

After concluding that there must be a "LOGOS" or divine "programmer" responsible for the specifications required for our universe, "from the cosmic constants to the DNA," cybernetics scientist David Foster states: "As to just 'where' LOGOS might be, we look for a place from which LOGOS can irradiate nature, and this would seem to be stars in general and (in our own case) the Sun."[15] Impressed by the sun's amazing catalytic abilities (that, against all odds, result in carbon and make life possible), and looking for the source of the yet more amazing specificity of DNA, Foster speculates that the sun is in the most logical place to send the DNA message to earth. "In that case," he says, "LOGOS would be something like 'the Sun God.'"[16]

In his immensely popular television program, *Cosmos*, Carl Sagan said, "Our ancestors worshipped the Sun, and they were far from foolish. It makes good sense to revere the Sun and the stars because we are their children."[17]

I do not mean to suggest that Sagan is advocating ritualistic sun worship today, but he is struggling with pagan metaphors to help us visualize his metaphysics; he is searching for ways to get us emotionally excited about the worship of impersonal works of creation (as did the ancient polytheists).

Sagan, of course, would not call them "works." They are collectively called "Cosmos," with a capital C, and his point is that "God" and Cosmos are one and the same. Many other scientists, feeling similarly compelled to make some acknowledgment of their findings of design (but unwilling to take this evidence to its logical end), routinely spell "nature" with a capital N and "universe" with a capital U.

As if to reprimand those who would one day abuse his scientific method (as Sagan does when he tries to use science to say that the Cosmos is all there is, was, or ever will be), Francis Bacon drew this important distinction: "Therein

Francis Bacon (1561-1626), father of the scientific method. —*Courtesy AIP Emilio Segrè Visual Archives* the heathen opinion differeth from the sacred truth; for they supposed the world to be the image of God. . . . But the Scriptures never vouchsafe to attribute to the world that honour, as to be the image of God, but only the work of His hands."[18]

I would argue that the scientific revolution began when Bacon and others demythologized *Nature* in order to study it as *nature,* the mere handiwork of God subject to His laws, not a mystical power free to act according to its own whims.

The laws themselves were established by God as His servants and are also subject to Him. Psalm 119, the chapter known for its exaltation of God's *moral* law, clearly puts God's *natural* law in its proper place: "You established the earth, and it endures. Your laws endure to this day, for all things serve you" (vv. 90b-91).

Sagan's task of motivating us to feel "loyalty"[19] to this deified Cosmos is difficult (if not perverse) in light of his doctrine that the universe is both impersonal and "indifferent to the concerns of such puny creatures as we."[20] The point of emphasizing all those "billions and billions" (popping the *b* each time he said it) of stars and galaxies was to show how insignificant we individuals are compared to the mystical Cosmos that gave us life.

It would appear that the alternatives to God—and the issues of heart and choice—have not substantially changed since the days of idol worship. The idea of "worshipping and serving created things" certainly includes both of the first two choices we mentioned (the universe and humankind as alternatives to God), and it is possible to choose both simultaneously. We work and live for material possessions, thinking thereby we best serve *ourselves.* In pantheistic religions, the line between worship of nature and worship of self is also blurred. And ancient polytheistic kings, when given the opportunity, often demanded worship, not only of their nature gods, but of themselves. In the same way, modern cosmologists may simultaneously honor both the universe and humanity as divine.

To some, it may seem absurd to think that scientists would consciously choose the *second* option, the option that *we* might be our own creator and designer. But remember, we have already described the PAP and FAP hypotheses, which say exactly that. Frank Tipler's *The Physics of Immortality* (a virtual bible of FAP doctrine) is immensely popular.

The panspermia hypothesis, the position taken by Sir Fred Hoyle, provides yet another way for us to be our own gods. Convinced from the microbiological evidence that life could not have evolved here on earth, Hoyle proposes the idea that it evolved elsewhere and traveled here in stages. Also convinced that the universe's design displays signs of tremendous intelligence, he writes that an intelligence which "was very big indeed" was involved in sending life here.[21]

Hoyle describes "microorganisms and genetic fragments . . . existing throughout space in prodigious numbers, riding everywhere on the light pressure of the stars," until "on Earth a creature at last arose with an inkling in its mind of what it really was, a whisper of its identity: *We* are the intelligence that preceded us in its new material representation—or rather, we are the re-emergence of that intelligence, the latest embodiment of its struggle for survival."[22]

Which Scientists?—God's Choice

On several occasions, people have asked me, logically enough, "If the evidence for divine design is so obvious to those who are intimately acquainted with it, why don't all these scientists put their faith in the God of the Bible? To which I now have the snappy reply, "We must attribute this to the fundamental goofiness of human nature." But more importantly, the question points to the fairness of a God who will not reward or punish His creatures according to their level of intelligence or skills in science. What kind of justice would reward us eternally according to our ability to do scientific research?

This often leads to the question, "Why wouldn't a loving God just bring *everyone* into His kingdom?" This question seems to be of such concern to both believers and unbelievers that we will give it special attention in the "Greatest Objections to the Gospel" chapter in Volume 3. For now let me simply point to a generally recognized truth: God must be more concerned with our hearts than with our heads.

True love requires mutual choice. The reader may not have suspected that all this talk of nature's design was going to lead to a discussion of true love, but it inevitably does.

If God desired to have an eternal, trusting relationship with a people who would willingly return His love, He could not simply create a people filled with love and trust, because that would not involve their wills. The very idea of a real *will to trust and love* requires the real possibility of a *will to distrust and reject*. Such acceptance or rejection can only come with real opportunity in life. God gives this opportunity, this call, to everyone; but He chooses for eternity only those who willingly accept and trust Him. How might He do this choosing? Should it be on the basis of intellectual achievement applied towards discovery of God? Should the choosing be done on the basis of finding those who are most intuitive in catching on to the plan—say the top two percent, since God is worthy of having only the very best?

What is needed to test faith is a clear *object* of faith. Love requires an object of love. And if, because of His grace, God wants to select as many as possible and make their willingness to love as easy as possible, what could better inspire their love than sacrificial love toward them first?

Obviously, a test of the will cannot involve inherited characteristics. If anything, there is an inverse relationship between earthly and heavenly advantages: "To whom much is given, much is required." For those who have been exposed to the fact that God so loved the world that He gave His only Son, more is required.

The fact that God's power and benevolence toward His creatures are also revealed to us in a more general way, through nature, should at least begin to create a softening of even the hardest hearts toward Him. But if science or logic could give us the full answers about God, we could take all the credit for finding Him. The story of the Bible is not the story of humanity's search for God; it is the story of how *God* has revealed Himself to *us*. It appears that He has given us science and logic just to whet our appetites, to drive us to His book.

Conclusion: Designed to Know Him

In our recent history, one by one we have discovered many of the laws of nature, but we have never discovered why there *are* laws of nature, why there's not chaos. Science has been unable to tell us anything about the cause of these laws. We only know that these laws result in order, in symmetry, in harmony, in balance, in mind-boggling, complex, highly-developed living forms that we can't duplicate in the laboratory with all our efforts, let alone expect them to happen by chance. The reason modern physicists are devoting their lives to finding a Grand Unified Theory (GUT) is that nature has taught them to expect symmetry, harmony, purposeful design, and simplicity at the base of all complexity—a single mind that ties it all together.

Though some popular scientist/celebrities and some textbook writers continue to speak of how the laws of science actually *govern* nature, the scientists themselves—especially those who have been the discoverers of these laws—have often been convinced that these perfect, purposeful laws point to a perfect, purposeful superintelligence. All the people quoted in these chapters dealing with design are well-known, highly respected astronomers, mathematicians, and physicists. None of those selected for quotes from this century (with the exception of two noncontroversial quotes from astronomer Hugh Ross) are Bible-believing Christians as far as I know; so this conclusion has not been prejudiced by the Christian viewpoint. Those who have rejected the God of the Bible have had to draw their conclusions in spite of the evidence, not because of it.

The acknowledged purposefulness and careful design observed in nature strongly suggest a God who is conscious, caring, and purposeful—in short, a *person*—not a purposeless, uncaring, unconscious force. As mentioned earlier, this certainly makes more sense than to think that personhood—the highest state of mind we know—is something that we possess but God lacks. Even without accepting the Bible's clear statement that we are made in His image, we can deduce that if the Creator's state of consciousness is different from ours, it must be a greater consciousness, not an inferior one. Much of the conjecture that God is less than a person stems from the old idea that God is one with the universe, an idea that is inconsistent with modern science (since the consensus of science is that the universe had a beginning and thus requires an outsider to begin it).

Belief in the "blind chance" position, it turns out, requires *considerable* faith. Such believers frequently put themselves in the difficult position of making claims of meaning in life without being able to give tangible reasons for it. Many have been influenced by the illogical sentiments of popular scientists whose statements they have apparently not thought through. The few blind chance advocates who speak of the pointlessness of existence should be commended for their consistency of thought. Taken to its logical end, such a view becomes so pathetic as to become almost funny; and so, though we cannot find quotes from many scientists who (like Steven Weinberg) admit to life's pointlessness without God, we can offer this thoughtful quote from an article entitled, *Adrift Alone in the Cosmos*—by Woody Allen:

> More than at any other time in history, mankind faces a crossroads. One path leads to despair and utter hopelessness. The other, to total extinction. Let us pray we have the wisdom to choose correctly. I speak, by the way, not with any sense of futility, but with a panicky conviction of the absolute meaninglessness of existence.[23]

At the end of the article, he explains why: "We have no spiritual center. We are adrift alone in the cosmos . . . ."[24]

Some might ask theists why their view should attach any more meaning to life than the blind chance position. Just because some supernatural Designer wound up the universe and let it go, should that fill our lives with meaning?

Believers in a careless, deistic god must concede the point. But if observations provide good evidence that the universe was not only designed, but designed with great precision and care, with *us* in mind, as today's physicists say,[25] then we are not alone in the cosmos. We are not beyond the Creator's thought and purpose. Rather, we are accountable to Him. We are bound by His care for us to ask that most human of questions: Who *is* my Creator? If God cares about us and made us with this desire to know Him, He should have provided a way. And if there is any chance that we can have a personal relationship with Him (as we suspect there must be*), the logical thing to do is to devote our short lives to finding the means, for this is the one relationship that can last, a relationship with something from outside of time, the one relationship that could give our lives *lasting* value.

Carl: As long as you're looking up at all the perfections in the heavens, it's easy to see the design and see how good God is. But sometimes, the way things go down here, well, I'm still left with many perplexing mysteries about life.

Fred: Such as?

Carl: Such as: If the universe is so perfectly designed, then why is it that people who snore always fall asleep first? And why is it that, when you ask someone to repeat what they just said, they always repeat just the part you already heard and leave out the one part you didn't?

Fred: These are very deep mysteries.

Carl: Or what about the problems of my nephew Ralph here? He's not so sure there's a good God, because he says life isn't fair. Ralph, tell us something about yourself.

Ralph: I lost my job, I lost my house, I lost my wife, I have an incurable disease, and I got mugged on the way here today.

Carl: At least your wife doesn't snore.

Fred: This brings up an important point about people who say they have logical problems with the gospel. Often there are more personal, deeper reasons.

Carl: My wife snores deeply.

Fred: I was thinking more about Ralph's problems.

Carl: Let's save Ralph's problems for Volume 2.

---

*Even Werner Heisenberg, chief founder of quantum mechanics, conceded that such a Soul to soul relationship ought to be possible. When asked, "Do you believe in a personal God?" Heisenberg answered: "May I rephrase your question? I myself should prefer the following formulation: 'Can you, or anyone else, reach the central order of things or events, whose existence seems beyond doubt, as directly as you can reach the soul of another human being?' I am using the term 'soul' quite deliberately so as not to be misunderstood. If you put your question like that, I would say yes."[26]

Fred: The real problem is how to reconcile two indisputable facts. Fact one is that the universe has been finely tuned, perfectly designed— it's "very good," as it says in Genesis 1. Fact two is that the earth is filled with suffering and violence. Even without a knowledge of fine-tuning from physics, most unbelievers somehow know that God must be perfect and righteous. Otherwise they wouldn't complain about how God sits idly by and allows all these evils in the world. They're right to wonder about how these two contrary ideas can both be true, because if they seek the answer they'll discover that God *has* done something about the problem of evil.

Carl: They'll discover what?

Ralph: About the problem of evil.

Carl: I heard that part.

Fred: *This is the most important implication of a perfectly designed universe. A good God would* **do something** *about all the evil in the world.* In the next chapter, we'll examine what His options might be.

Shocking Summary Statements and Stimulating Conversation Starters

The most important thing we should note about the evidence for design is that it brings good news. Life is not pointless. Consider the alternative to God. If the universe is an accident, then our lives do not have meaning or lasting value.

Only something that lasts can give our lives lasting value. Also, because all value is based on relationships (between one thing and something outside itself), our universe can have value only when there is something outside it to value it.

In the same way, objective moral values point to an outsider. While values may change from culture to culture, objective values do not. If the blind chance advocate acknowledges that society was obviously wrong when it upheld racism, slavery, or human sacrifice, then we must ask, *Wrong* relative to what? By using that word, the materialist acknowledges that there must be a higher standard by which to judge.

Design drives scientists to a Designer. Sir James Jeans, whose formulas led to our modern theories of galaxy formation, wrote, "Mind no longer appears as an accidental intruder into the realm of matter; we are beginning to suspect that we ought rather to hail it as the creator and governor of the realm of matter." David Foster says: "I have found few, from Einstein to Schrödinger, who at some stage or another did not have to introduce God."

In the matter of deciding who's running the universe, we all have three options: the universe itself, mankind, or God. Because this decision will have obvious repercussions on the matter of who we will allow to run our lives, we cannot make this decision on the basis of logic alone. We are not dispassionate intellects; we are willful creatures.

Why wouldn't a loving God just bring everyone into His kingdom? True love requires mutual choice. If God desired to have an eternal, trusting relationship with a people who would willingly return His love, He could not simply create a people filled with love and trust, because that would not involve their wills. The very idea of a real will to trust and love requires the real possibility of a person's will being used to distrust and reject.

What is needed to test faith is a clear object of faith. Love requires an object of love. And if, because of His grace, God wants to select as many people as possible and make their willingness to love as easy as possible, what could better inspire love than sacrificial love toward them first?

Once we understand that the universe has been perfectly designed, we see how far short our own world, filled with suffering and violence, falls from perfection. The most important implication of a perfectly designed universe is that a perfect Designer would *do* something about the problem of evil in our world.

Notes for Chapter 11

1. Carl Sagan, *Cosmos* (New York: Random House, 1980), p. 127.

2. Ibid. Emphasis added.

3. Ibid., p. 345.

4. Ibid.

5. Sagan, *Cosmos* (the television series), Episode 4.

6. Steven Weinberg, cited by Freeman Dyson, *Disturbing the Universe* (New York: Harper & Row, 1979), p. 246.

7. Colin A. Russell, *Cross-currents—Interactions Between Science and Faith* (Grand Rapids: William B. Eerdmans Publishing Company, 1985), p. 174.

8. Steven Weinberg, *Dreams of a Final Theory—The Search for the Fundamental Laws of Nature* (New York: Pantheon Books, 1992), p. 16.

9. John D. Barrow and Joseph Silk, *The Left Hand of Creation—The Origin and Evolution of the Expanding Universe* (New York: Basic Books, Inc., Publishers, 1983), p. 26.

10. Barry Parker, *Creation—The Story of the Origin and Evolution of the Universe* (New York and London: Plenum Press, 1988), p. 282.

11. James Jeans, *The Mysterious Universe,* cited by A.E. Wilder Smith, *Man's Origin, Man's Destiny* (Wheaton, IL: Harold Shaw Publishers, 1970), p. 9.

12. David Foster, *The Philosophical Scientists* (New York: Dorset Press, 1991), p. 169.

13. Ibid., p. 172.

14. Ibid., p. 179.

15. Ibid., p. 179.

16. Ibid., pp. 158-161.

17. Carl Sagan, cited by Howard J. Van Till, Davis A. Young, and Clarence Menninga, *Science Held Hostage—What's Wrong with Creation Science AND Evolutionism* (Downers Grove, IL: InterVarsity Press, 1988), p. 159.

18. Francis Bacon, cited by Colin A. Russell, *Cross-currents—Interactions Between Science and Faith* (Grand Rapids: William B. Eerdmans Publishing Company, 1985), p. 76.

19. Carl Sagan, *Cosmos* (New York: Random House, 1980), p. 345.

20. Ibid., p. 250.

21. Fred Hoyle, *The Intelligent Universe,* (New York: Holt, Rinehart and Winston, 1983), p. 161.

22. Ibid., p. 239.

23. Woody Allen, "Adrift Alone in the Cosmos," *New York Times* (August 10, 1979).

24. Ibid.

25. Freeman Dyson, *Disturbing the Universe* (New York: Harper & Row, 1979), p. 250.

26. Werner Heisenberg, "Positivism, Metaphysics, and Religion," *The World Treasury of Physics, Astronomy, and Mathematics,* ed. by Timothy Ferris, (Boston: Little, Brown and Company, 1991), p. 826.

PART V

Which Cosmic History Is Ours?

An Apollo view of earth from 1/4 million miles away, showing the west coast of
North America. —*Courtesy of NASA*

CHAPTER 12

Is the Gospel Logical?

*Faith . . . is the art of holding on to things your reason has
once accepted, in spite of your changing moods.*

—C.S. Lewis

What I am saying is true and reasonable.

—The Apostle Paul

The Logical Cosmic Options:

There is good logic in believing the one cosmic history that fits what we know of God.

When I speak here of what we know of God, I'm not referring to what we have heard from human tradition, but what we must infer from natural revelation. This book describes the observations which have led scientists today to conclude that a super-intellect must have carefully selected the laws, within very narrow parameters, that are necessary to make our universe—and human life—possible.

If the Creator is the sustaining, caring God that our universe shows Him to be, then we should logically expect Him to choose a sustaining, caring alternative from among any options He might have in dealing with us.

Once we understand that the universe requires a super-intelligent first cause, we can consider seven basic alternatives—seven cosmic histories that *might* have been—given what we know from *this* universe. The number might be greater or smaller if we wished to divide the alternatives differently, but we must notice that, except for the last one, they are all bad news for us. Only one option points to the wise, caring, sustaining God whose universe we now observe.

Cosmic History No. 1

God created nothing. He just enjoyed *Himself* forever.

Cosmic History No. 2

God created a beautiful universe, sustained it by harmoniously-working laws, but refrained from creating life—or at most only plant and lower forms of animal life. He appreciated it as a master artist forever, with no persons to show it to or share it with.

Cosmic History No. 3

God created all this plus self-conscious, free-willed beings like Himself: *persons.* Having been given the freedom of choice, the persons eventually declared their independence from God, *broke* His harmoniously-working laws which were necessary for the good life God planned for them, the situation became very unpleasant, and God exterminated them. The end. An experiment that failed.

Cosmic History No. 4

God created self-conscious beings, persons like himself, except He programmed their wills so that they were limited to doing *His* will, so that they were *incapable* of messing up His universe. Both God and His persons lived happily ever after, except that the persons served God out of compulsion—not love. Without choice. Serving like zombies under a spell. Notice that up until this point, these cosmic histories are clearly contrary to our own—we know these histories are false.

Cosmic History No. 5

God created persons in His own image, the persons broke His harmoniously-working laws, the situation became very unpleasant, and so God *left* the whole mess, going off to other dimensions and letting the persons create their own hell until they either exterminated themselves or the messed-up universe eventually fell apart. Many people feel this last one describes *our* universe—the idea that God just wound up the universe and let it go. However, such an idea is not consistent with an all-knowing, caring God. The God whose carefully designed universe we observe would not have started something He couldn't care for.

Cosmic History No. 6

God created persons with freedom of choice. Everything was just as we see it except that when the persons started breaking His moral laws, God simply forgave them. God bent His laws of justice to accommodate all the hate and violence that people inflicted upon one another. He overlooked their little problems with racism and their various injustices. And so, throughout all eternity, God provided the rain and the sun and everything necessary to sustain the greed and the wars and all the suffering that resulted.

This is some folks' idea of the gospel. We just do our best and God just forgives. But without God actually *doing* something about the injustice, this is still not the best alternative—or even an option that shows forethought or care.

Cosmic History No. 7

By elimination, we finally come to something that looks just like the good news we find in the Bible. We have all the bad news described above, except that, when the persons broke the harmony of His universe, God had a plan. He didn't exterminate His persons and He didn't just overlook their crimes; He didn't run out on them and He didn't *force* them to fit into the harmony of His universe.

So here's what God did. He looked upon these persons who were responsible for all the death and destruction their lawbreaking had caused—and He pronounced a just death sentence on them. That's the bad news. But He also granted a full pardon to all who would take it. That's the gospel. In Volume 4, we will explain more fully how God could justly implement both a death sentence and a pardon: only by getting involved Himself, only by means of a God-man. He had to be a mortal person, because God cannot die, and He had to be God, because when a human dies, he suffers the just payment for his own sins; only a sinless person could possibly pay for the sins of another.

Our part in all this is merely to choose His way over our way. That's faith. We come to faith when we humbly acknowledge that our way is wrong. His way is right. The Bible calls this "repentance." And on the basis of our faith in Jesus Christ, God eventually changes us, allowing us to have abundant life in harmony with His laws, making us fit for His company throughout eternity. Only now, instead of *forcing* us to love Him, He has given us the freedom to choose Him or to reject Him. After experiencing first-hand the despair that comes from following our own misguided path, we can either appreciate God's way more deeply, or we can stubbornly decide to keep going on our own to see where our way finally leads us.

The apostle John wrote of the new type of persons that come into being when they receive God's source of life and light, describing them as "born again" or "born of God":

> He was in the world, and though the world was made through Him, the world did not recognize Him. He came to that which was His own, but His own did not receive Him. Yet to all who received Him, to those who

believed in His name, He gave the right to become children of God—children born not of natural descent, nor of human decision or a husband's will, but born of God (John 1:10-13).

A Logical Examination of History:

There is good logic in accepting the one message, of all those claiming to be from God, that is grounded in history.

Of course, other religions have their historical figures. But the gospel is centered, not just on the *teachings* of an historical person, but on what that person *did*. Take away whatever Buddha did in history and you can still follow the Noble Eightfold Path. But take away what *Jesus* did—take away His death and resurrection—and the Bible says our faith is futile:

> And if Christ has not been raised, our preaching is useless and so is your faith. More than that, we are then found to be false witnesses about God, for we have testified about God that He raised Christ from the dead. . . . And if Christ has not been raised, your faith is futile; you are still in your sins. Then those also who have fallen asleep in Christ are lost. If only for this life we have hope in Christ, we are to be pitied more than all men (1 Corinthians 15:14-15,17-19).

The gospel hangs on two historical events: Jesus' death by crucifixion, which no one seriously disputes, and His resurrection from the dead. Faith enters the picture when we consider whether Jesus actually rose from the dead—not because this event is illogical, but because whether we accept this event and all its implications is a matter of *choice*. If Jesus is who He claimed to be, then it is not unreasonable that He should have the power and the reason to rise from the dead. In fact, it would be unreasonable for God to be bound by death.

No other religious leader ever claimed to be the Savior of the world, to be one with the Creator/God in a unique way, to personally have the power to take away our sins, or to be capable of raising Himself from the dead. As for His reason, the Bible tells us His resurrection serves as proof of His divine claims and of His power to raise *us* from the dead also (Acts 26:23). Jesus' identity is bound up inextricably with His resurrection. That's why He said, "I am the resurrection and the life" (John 11:25a).

There *is* no gospel without the historical events. Of all the religions, no other has left such evidence of God's involvement with humanity (see Volume 4 for the historical and archaeological evidence). If ever God reached into our world, it was through a human named Jesus.

Logically Addressing the Objections*
Objection 1: "Faith Excludes Logic"

The Objection

Because of their misunderstanding of the biblical doctrine of faith, some people argue that the gospel isn't supposed to be logical—it's just supposed to be believed. For some, it is a mark of spirituality to be able to throw all human reason to the wind in order to have faith. After all, aren't we saved by faith alone? Isn't it impossible to *reason* our way to heaven? Doesn't the Bible itself say that the gospel is foolish?

The expressions "leap of faith" and "blind faith" obviously leave no room for logic. Unbelievers are the most frequent users of these expressions to describe their perception of biblical faith.

The Response

While it is very true that the Bible describes faith as the conviction of things we can't *see* (Hebrews 11:1), it never says that faith is the conviction of unreasonable things that our minds can't *comprehend*. Science gives us many facts that we know though we cannot see them; this does not make those scientific convictions illogical. And the Bible does not say that faith is the conviction of things that contradict what we *can* see, or that faith is the *feeling* that assures us of something we merely *wish* to believe. C.S. Lewis's definition of faith, quoted at the beginning of this chapter, makes the point, as does his famous statement about seeking truth first, not comfort:

> In religion, as in war and everything else, comfort is the one thing you cannot get by looking for it. If you look for truth, you may find comfort in the end: If you look for comfort you will not get either comfort or truth—only soft soap and wishful thinking to begin with and, in the end, despair.[1]

Speaking of how Jesus fulfilled the Hebrew prophecy that the Messiah would rise from the dead, Paul told the Roman governor Festus, "What I am saying is true and reasonable" (Acts 26:25). He admonished the Corinthian believers to "stop thinking like children. In regard to evil be infants, but in your thinking be adults" (1 Corinthians 14:20). He doesn't ask his readers to lower their standards of sense or judgment in order to believe him, but rather tells them: "I speak to sensible people; judge for yourselves what I say" (1 Corinthians 10:14).

---

* For a more detailed treatment of the greatest objections to the gospel, see the last chapter of Volume 3: "Hell, Hypocrites, Heathens, and Harps in the Hereafter."

Peter urges his readers to use "sound judgment" (1 Peter 4:7, NASB). Luke commends the "noble" Bereans who "examined the Scriptures every day to see if what Paul said was true" (Acts 17:11). The New Testament writers fill their letters and treatises with sound, logical arguments that are intended to convince reasonable people, not challenge their credulity. Any unreasonable or blind leap of faith is not the kind the Bible describes.

1 Corinthians 1:18 does tell us that the message of the gospel is foolishness to many who reject it—but a few verses later we're told that to others it displays the very wisdom of God. In fact, the passage is a good example of how the Bible takes up the cause of healthy skepticism. Like the true skeptic, the Bible always displays a leery attitude toward the fashionable, so-called "wisdom" of any particular philosophy or age.

To say that we are saved by faith alone does not mean that faith cannot be logical. We would not claim that we are saved by faith plus logic any more than we would say that we're saved by faith plus good deeds; but if a faith that doesn't show itself by its deeds is a dead faith (James 2:26), we might at least wonder if a faith that doesn't engage the mind is, to say the least, missing something. God wants our love toward Him to involve not only our hearts and souls, but also our *minds* (Matthew 22:37).

Last, it is also true that logic alone cannot bring us to a full knowledge of God. Natural revelation can take us only so far—just far enough to point us strongly to special revelation. As Chapter 9 showed, the evidence of caring, purposeful design in the universe gives us powerful reasons to seek out God's special communication in His Word. Those who do will then need to continue to use their faculties of reason in order to understand and apply the Bible's message. The use of such reasoning faculties does not inhibit the work of the Spirit in our lives—we do not throw away all logic and lose all self-control each time we pray; rather, we are enjoined to "be clear minded and self-controlled so that you can pray" (1 Peter 4:7).

Objection 2: "The Gospel Sounds Too Foolish"

A Skeptic's Objection

The core ideas of the Christian gospel cannot be taken seriously in the 20th and 21st centuries—the idea of God becoming human, the concept of one person dying for the sins of another, and for that matter the whole concept of sin. They certainly aren't taken seriously by most in academic circles. These concepts may have met the need a thousand years ago, but now they're outdated. They are not sophisticated enough for people today and, if Christianity is to survive, should be adjusted to fit today's mores.

A Believer's Response

The gospel does not pretend to be sophisticated. Should the way to eternal life be something that only the intellectually gifted can perceive? Should it be the reward for those who have reached the highest levels of education? Or should it be something that even a child can understand? In Luke 10:21, Jesus prayed, "I praise you, Father, Lord of heaven and earth, because you have hidden these things from the wise and learned, and revealed them to little children."

If the infinite Creator of the universe wanted to communicate to *all* us little earthlings what He's like, how could He show us more clearly than by becoming one of us? If He wanted to communicate to us the seriousness of breaking His moral law, how could He show us more forcefully than by demanding that the most valuable, worthy thing in the universe—His own life—be forfeited as a penalty? And if He wanted to tell us how much He *loves* us, how could He do it more dramatically than by dying for us? Jesus said, "Greater love has no one than this, that he lay down his life for his friends" (John 15:13).

As for the idea of changing God's message in order to conform to the changing moral standards of society—is this really logical? Who should conform to whom? Has anything about human *nature* changed in the last few thousand years, so that we have reason to believe that our changing mores have improved the human condition? Have we stopped locking our doors at night? Has crime in the cities decreased? Do the nightly news stories fill us with hope? Is the problem of sin less relevant today?

As we mentioned, the Bible shows healthy skepticism toward the changing fashions of human philosophy. 1 Corinthians 1:20 says, "Where is the wise man? Where is the scholar? Where is the philosopher of this age? Has not God made foolish the wisdom of the world?"

The gospel was not any more fashionable when it was first preached than it is today. It was a problem for the Jews who were expecting a triumphant, political Messiah, and it was foolish to Gentiles who believed that no reputable Savior would be crucified (1 Corinthians 1:22-23). And yet, amazingly, these attitudes did not stop the gospel from spreading. The apostles didn't change the gospel into something that would conform to the wisdom of their day; Paul summed up the gospel he preached in two, foolish-sounding words: "Christ crucified." And he called the crucified Messiah the power of God and the wisdom of God—not because the message was so well received by all, but because God enabled many Jews and Gentiles to receive it in spite of their biases.

Objection 3: "The Gospel Is Too Narrow"

A Skeptic's Argument

The United States and Europe are increasingly becoming lands of great cultural and religious diversity. The world, of course, is populated by people of many faiths. How can we take the gospel seriously in light of its exclusive claims? In John 14:6, Jesus said, "No one comes to the Father except through me." In Acts 4:12, Peter said, "Salvation is found in no one else, for there is no other name under heaven given to men by which we must be saved." How can that be good news for people who decide to follow other religions or to follow God in their own way and just be good people? The gospel is too narrow for our world today. There should be many ways to heaven or nirvana or whatever.

A Believer's Response

The end of this argument displays its logical flaw, for it is not possible for the same person to go to hell and come back reincarnated as a cow at the same time. Nor is it logical that one good person should go to heaven and another, having lived the same way, should lose all personal identity and go to nirvana. The gospel is narrow; but the very nature of *truth* is narrow, according to our laws of logic, which tell us that two contrary statements cannot both be true.

Everything we know from nature about reality teaches us that there is one set of laws for everyone; e.g., people cannot pick and choose from among various laws of gravity that will apply to them, according to their preferences.

The end of Volume 3 contains more thoughts on people who have had little opportunity to hear the gospel, but strictly from a logical viewpoint, it makes perfect sense that God would establish one way to save all people. Unlike blind faith, reasonable faith must have an object. If people could just believe in the power of pickles and green cheese to bring them to heaven—or anything else that seemed attractive to them—there would be *no* opportunity for people to exercise *faith* in God and in His way.

And this leads us to our final point:

A Logical Application:
There Is Good Logic in Making the Gospel a Personal Thing

In science, logic alone can take us just so far. No discovery of science was ever made by logic alone. Logic does not tell us how to make inferences; the human mind must surmise and choose. Logic assists us in eliminating bad inferences. In the same way, in matters of faith, logic can lead us to the very threshold of faith, but then it is up to the human spirit to make a personal choice.

Today in Western societies, it has become fashionable to say that religion should be "a personal matter." Technically, if by religion one means his practice of duties for God, I would agree; it is wrong to broadcast one's good deeds as the hypocrites do. But I am not talking here about religion— I'm talking about a relationship with the person of Jesus Christ. If by "making the gospel a personal thing," one means that he intends to keep word of this relationship to himself and share it with no one else, then it would not be logical to call the gospel "good news."

By "making the gospel a personal thing," I mean that we must personally accept Jesus Christ in order to appropriate the good news for ourselves. We must have a personal relationship with God, not just with our church. We must have a relationship that gives us the right to talk *to* Him, not just *about* Him. We must take advantage of the opportunity that even quantum physicist Werner Heisenberg said should be possible: to reach God "as directly as you can reach the soul of another human being."[2]

But how can we reasonably expect to do this? First we must be clear that the *doing* of everything that has to be done to obtain this eternal relationship has already been done. When Jesus started preaching the gospel, He said the kingdom of God was coming to *us*, not that we had to do something to get to the kingdom of God. But in order for each of us to personally accept the gospel, Jesus did give us two, simple imperatives: repent . . . and believe. In Mark 1:15, Jesus said our response to the gospel should be to "Repent and believe the good news!"

Personal Repentance Is Logical

How do we show God that we are fit for His company throughout all eternity? What can we do to be just before a holy God? Through a parable, Jesus explained that there are two ways to be justified in God's sight: the right way . . . and the wrong way:

> Two men went up to the temple to pray, one a Pharisee and the other a tax collector. The Pharisee stood up and prayed about himself: "God, I thank you that I am not like other men—robbers, evildoers, adulterers— or even like this tax collector. I fast twice a week and give a tenth of all I get." But the tax collector stood at a distance. He would not even look up to heaven, but beat his breast and said, "God, have mercy on me, a sinner." I tell you that this man, rather than the other, went home justified before God (Luke 18:10-14a).

Like the first man, we can justify ourselves by thinking of all the people who are worse than we are. We can tell ourselves and God that we're not so bad and even do good deeds to prove it. This method succeeds in justifying ourselves to ourselves, but not to God.

To repent we must acknowledge our sinfulness before a holy God, understanding that, except for His mercy, our condition is hopeless. We might acknowledge to Him that we've done our share to mess up a perfectly good universe. Repentance should express a desire to *turn* from going our own way—since that old way leads nowhere except to death—and a desire to let Him lead us, His way, to life as He intended it to be.

Taking the First Step of Faith Is Logical

A reasonable faith needs an object, and God's Word clearly points us to Jesus Christ, the One who has met all of God's just terms of punishment on our behalf, through His death on the cross. The giving of His only Son's life for us was God's communication of both His great love and perfect justice—but it was more than just an object lesson. It was God's means of turning sick persons into whole persons, dying persons into eternal persons who can experience all the abundant life that God intended for them to enjoy in fellowship with Him.

Biblical faith is more than intellectual belief; it is a logical *choice* of God's way over our way. Anyone wishing to make sure that he has taken this first step of faith should confess his moral failings to God and then thank Him, in his own words, for sending Jesus to die for him. From that point on, we express faith by daily trusting Christ to lead us and be Lord of our lives.

Getting Rooted in Christ Is Logical

It is not God's intention that we leave the Bible to the theological "experts." Paul says that his purpose for all the Christians he could reach was that "they may have the full riches of complete understanding, in order that they may know the mystery of God, namely, Christ, in whom are hidden all the treasures of wisdom and knowledge" (Colossians 2:2-3). He instructs the people of Colosse to get "rooted and built up in Him [Christ]" (verse 6).

Getting rooted and built up in Christ requires regular time spent alone with the Lord in prayer and Bible-reading. There is nothing reasonable about a relationship in which there is no two-way communication. True faith will not be satisfied with salvation as a kind of "fire insurance"; rather, faith urges us on to greater curiosity about our faith's Author and Perfecter, to a greater understanding of His will for us, to greater ways of expressing our love for Him, to greater compassion for all the people He brings into our lives. His Word also instructs us to meet regularly with other Christians for mutual encouragement, teaching from the Bible, and worship (Hebrews 10:24-25, Acts 2:42).

One word of caution though: Not every church takes seriously its responsibility to teach the Bible. It is our responsibility to do some skeptical investigating in order to find an assembly of believers that puts the Word of God above the words and traditions of people.

Conclusion

So, for all the reasons just set forth, we conclude that personally accepting Jesus Christ and His gospel is a supremely reasonable move for anyone who is interested in emotional well-being, in abundant and eternal life.

There is good logic in personally accepting the one plan that frees us to serve God out of love instead of from fear. Once we realize that we have to depend completely on Jesus and that we can't *earn* eternal life, then our good deeds can be acts of love, not acts of desperately trying to score points with God. The good news frees us to serve God out of gratitude, not from compulsion or fear of being condemned. Then we're free to truly *love* God, with all our hearts, souls, and minds. To be free of the feeling that we must constantly strive to establish our own worth is a freedom we'll never know without Christ.

Shocking Summary Statements
and
Stimulating Conversation Starters

 There is good logic in believing the one cosmic history that fits what we know of God. One thing natural revelation tells us about God is that He has carefully selected the laws of the universe, within very narrow parameters, that are necessary to make human life possible. We should logically expect that such a sustaining, super-intelligent, and caring Creator should choose a sustaining, caring, and workable alternative from any options He might have in dealing with us. Of the seven cosmic options, only one shows intelligent forethought or care.

The one cosmic option that shows forethought or care is that God, if He bothers to make self-conscious beings at all, makes them with the ability to accept His purposes or to reject them. If these beings choose to make a mess of their universe, then, according to this cosmic option, God is ready with a plan. He does not exterminate them; nor does He simply overlook their injustices; He does not run out on them, and He doesn't *force* them to fit into the harmony of his universe. Rather, His solution involves giving them a choice to be changed; it involves a just death penalty and a full pardon, perfect justice and unbounded mercy. It offers people the best possible reason to enter a love relationship, by demonstrating sacrificial love toward them first.

There is good logic in accepting the one message, of all those claiming to be from God, that is grounded in history. The gospel is centered, not just on the teachings of an historical person, but on what that person *did*. The gospel hangs on two historical events: Jesus's death on a Roman cross, which no one seriously disputes, and His resurrection from the dead. Faith enters the picture when we consider the latter—not because this event is illogical, but because whether we accept this event and all its implications is a matter of choice. If Jesus is who He claimed to be, then it is not unreasonable that He should have the power and the reason to rise from the dead.

If ever God reached into our world, it was through Jesus. No other religious leader ever claimed to be the Savior of the world, to be one with the Creator in a unique way, to personally have the power to take away our sins, or to be capable of raising Himself from the dead. Whether we are judging by the profusion of historical material written about Him, or by the changed lives and the abundance of hospitals and charities in His name, we must observe that no other religion and no other person has left such evidence of God's involvement with humanity.

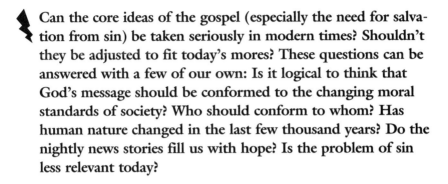 Can the core ideas of the gospel (especially the need for salvation from sin) be taken seriously in modern times? Shouldn't they be adjusted to fit today's mores? These questions can be answered with a few of our own: Is it logical to think that God's message should be conformed to the changing moral standards of society? Who should conform to whom? Has human nature changed in the last few thousand years? Do the nightly news stories fill us with hope? Is the problem of sin less relevant today?

But what about the idea of God becoming man? What about the concept of sacrifice? Aren't these concepts primitive? By no means. If the Creator of the universe wanted to communicate to us (moderns and ancients both) what He is like, how could He show us more clearly than by becoming one of us? If He wanted to communicate to us the seriousness of breaking His moral law, how could He show us more forcefully than by demanding that the most valuable thing in the universe be forfeited as a penalty? And if He wanted to tell us how much He loves us, how could He do so more dramatically than by dying for us?

But isn't the gospel too narrow for our world today? Shouldn't there be many ways to heaven or nirvana or whatever? Such questions are logically flawed, for it is not possible for the same person to go to heaven and come back reincarnated as a cow at the same time. The gospel is narrow, but the very nature of *truth* is narrow. Reasonable faith must have an object. If people were free to pick and choose from among a variety of contradictory belief systems, there would be no opportunity for them to exercise *faith* in God and in His way.

It is not logical to try to justify ourselves by thinking of all the people who are worse than we are. This method may succeed in justifying ourselves to ourselves, but not to God.

There is good logic in making the gospel a personal thing. Biblical faith is more than intellectual belief; it is a logical *choice* of God's way over our way. We must *personally* accept

Jesus Christ as Savior in order to appropriate the good news for ourselves. We must have a personal relationship with God, not just with our church. We must have a relationship that gives us the right to talk *to* Him, not just *about* Him.

 True repentance involves turning from our old ways; true faith involves giving, not just lip service, but our hearts and lives to Him. If this seems too difficult for us, we should understand that it *is.* Once we understand that none of this is possible in our own power, once we realize that we have to depend completely on Jesus and that we can't *earn* eternal life, then our good deeds can be acts of love, not acts of desperately trying to score points with God. The good news frees us to serve Him out of gratitude, not from compulsion.

Notes for Chapter 12

1. C.S. Lewis, *Mere Christianity* (New York: The MacMillan Company, 1943, 1972), p. 39.

2. Werner Heisenberg, "Positivism, Metaphysics, and Religion," *The World Treasury of Physics, Astronomy, and Mathematics,* ed. by Timothy Ferris (Boston: Little, Brown and Company, 1991), p. 826.

Johannes Kepler (1571-1630). "A true bigot, a true believer. . . . This passionate belief turned out to be right" (Arno Penzias commenting on Kepler's leadership in the scientific revolution). —*Courtesy of the Mary Lea Shane Archives of the Lick Observatory*

What Everyone Ought to Know About the Origin of Science

Real scientists don't believe the Bible, right? Especially the world's all-time greatest scientists, right? Modern science began when certain intellectuals turned from spiritual beliefs to a belief that the physical universe is all there is, right? Wrong.

At a time when we pay such homage to science and modern technology, we do well to be reminded that, far from being hindered by Bible-believers, science has been led by them. The very basis upon which we approach knowledge "scientifically"—the scientific method that many assume is opposed to faith—was developed by people who readily professed Christ as their personal Savior. Most people today are unaware that the majority of pioneers of science were outspoken Christians—not because there was ever a conspiracy among the writers of science textbooks to keep this quiet, but simply because the personal beliefs of these scientists were irrelevant, from the educators' viewpoints.

So what possible connection could there be between an ancient, non-scientific book like the Bible and the beginnings of modern science? What does the Bible have to offer the modern scientist?

A Love for Truth, Not Tradition

The scientific revolution was not ushered in during a movement toward atheism, but during a reformation of faith. The Reformers were concerned with truth, not just tradition; with a personal relationship, not just a handed-down religion. This turn from tradition to personal experience in the spiritual realm encouraged a new self-reliance and a re-evaluation of old thinking in every field. And we will find a Bible believer at the foundation of almost every field of scientific knowledge, often a believer who was strong enough in his faith to persist in it in spite of opposition from the establishment.

Frequently persecuted as nonconformists, dissenters, or heretics by the established church, many of these people of faith were living illustrations of the difference between a relationship with an institution and a relationship with Jesus Christ. Whether they were Calvinists running from ecclesiastical authority or Puritans excluded from cultural privileges by the Church of England, the "natural philosophers" who were most earnest in their pursuit of a personal relationship with Christ were the very ones who were likeliest to make great contributions to science.

Although these nonconformists were excluded from the universities in England because of their dissent from the Anglican church, their dissenting academies for many years exposed their children to more science teaching than anywhere else in England. The moral courage and independent thinking that they displayed in matters of faith were the very traits that proved so vital in the mission of science.

Ancient dualistic philosophers like Plato had left their mark on the West, influencing many to believe that the physical world was evil while only the spiritual was good. According to such ideas, holiness was gained by withdrawing from the world; such views certainly did nothing to encourage investigation of the physical universe. Those who came under a biblical influence, however, learned that creation was declared to be "very good," and that humans were given a divine mandate to "subdue the earth."

Many today, of course, have been told only of the apparent conflicts between science and the Bible. What about the trouble Galileo got himself into when he proclaimed that the sun and planets revolved around the earth? This most popular example actually helps to make our case. It does not demonstrate conflict between science and the Bible, but between science and tradition. The established church took this tradition from the Greeks, not from the Bible.

Biblical Skepticism

Though the Bible was not written to teach science, it does contain principles that have prepared and encouraged scientists to excel in their work. The Bible encouraged the fifty discoverers in this section to strive for wisdom, truth, knowledge, and understanding, to test and examine all things, to study the works of God. It discouraged them from being gullible, from being conformed to the world, from being deceived by traditions, by authorities, or by their own hearts.

When common people began for the first time to read the Bible in their own languages, they could not help but to be influenced by the Scriptural emphasis on inductive reasoning: on proving, testing, examining, tasting. Superstition, astrology, etc. were specifically condemned; nature was de-mythologized and brought back to its rightful place as the handiwork of God with no power of its own. Bible readers also learned to be realistic about the fallibility of human authorities. Biblical skepticism drove our fifty discoverers to question things that all the world had believed were settled issues.

Training in Humility, Diligence, and Honesty

Further, the biblical traits of humility, diligence, and honesty (among others) are necessary for good science. A lack of these has caused some scientists to fudge their data or to publish their work prematurely to gain recognition. Others have kept important discoveries secret in order to be sure no one else could build upon their work and gain credit before them. Christians who displayed humility and honesty have published their findings in a timely way, giving others a chance to build upon their work.

Personal priorities concerning money caused some to devote all their time to research and continue making great discoveries (often while forgoing profitable opportunities in industry, as in Carver's and Faraday's cases). Many exhibited diligence in their work that had a lot more to do with learned, biblical character traits than with innate genius. Some, like Kepler or Morse, specifically attributed their ability to persevere in spite of countless setbacks to their faith in a God of purpose.

The great discoverers in science demonstrated that honesty is thinking's prerequisite. We can't think true thoughts and come up with true conclusions if we start out with a predisposition to accept only what we want to accept. An honest scientist cannot simply read into the data what he wants to see; he must re-evaluate all his presuppositions in light of the data. Even while criticizing certain sub-groups among Christians today, Harvard paleontologist Stephen Jay Gould praises the work of "great scientist-theologians of past centuries." Gould writes that their work was distinguished by "a willingness to abandon preferred hypotheses in the face of geological evidence. They were scientists *and* religious leaders . . . ."*

---

*Stephen Jay Gould, "Creationism: Genesis vs. Geology," in *The Flood Myth,* ed. by Alan Dundes (Berkeley, CA: University of California Press, 1988), p. 434.

Belief in a Creation of Order and Harmony

The one prejudice these pioneers shared has now become the one bias accepted by modern science: a predisposition to look for simplicity and order in nature. Of course, this conviction belonged not only to the dissenters, but often to members of the established churches as well; thus our list below includes many of them who also displayed a special dedication to their Lord.

The clear thinking behind these pioneers' achievements was often the result, not only of an attitude of nonconformity to the world, but of a conviction that God has created a universe of purposeful order and balance. Time and again, this presupposition has proved to be correct. Today no one can lightly dismiss these Christian thinkers as unreasoning fanatics, especially when we realize that we are benefiting from their discoveries every time we flip an electrical switch, type into a computer, submit to modern medicine, or watch television.

In my interview with Nobel prize-winning physicist Arno Penzias, he spoke of how all our expectations of order and simplicity in nature go back to the triumph of Kepler. Kepler, as a "true believer," sought a way to reconcile the lawgiving, orderly God of the Bible with the natural motions of the universe. Planets that moved in irregular, senseless patterns didn't fit the Bible's picture.

Heeren: What do you feel makes physicists think there must be some sort of ultimate, short explanation for all the laws of nature? Is it just their desire to find explanations for everything, or is there good reason to expect a unified theory?

Penzias: That really goes back to the triumph, not of Copernicus, but really the triumph of Kepler. That's because, after all, the notion of epicycles* and so forth goes back to days when scientists were swapping opinions. All this went along until we had a true bigot, a true believer, and this was Kepler. Kepler, after all, was the Old Testament Christian. Right? He really believed in God the Lawgiver. And so he demanded that the same God who spoke in single words and created the universe is not going to have a universe which has 35 epicycles in it. And he said there's got to be something simpler and more powerful.

---

* Epicycles were proposed by the second-century astronomer Ptolemy to account for the irregularities in planetary motion, which are now accounted for by Kepler's discovery that the planets orbit the sun in an ellipse with the sun at one of the two foci. To Ptolemy, however, each planet appeared to move in an epicycle, a circle whose center moves around the circumference of a larger circle.

Now he was lucky or maybe there was something deeper, but Kepler's faith was rewarded with his laws of nature. And so from that day on, it's been an awful struggle, but over long centuries, we find that very simple laws of nature actually do apply. And so that expectation is still with scientists. And it comes essentially from Kepler, and Kepler got it out of his belief in the Bible, as far as I can tell. This passionate belief turned out to be right. And he gave us his laws of motion, the first real laws of nature we ever had. And so nature turned out to redeem the expectations he had based on his faith. And scientists have adopted Kepler's faith, without the cause.

Scientists Who Adopt the Cause Too

How vocal should a scientist be about his spiritual beliefs? Is it ever permissible for a scientist to say that a scientific finding supports the Bible? The discoverers whose stories are told below may have differed with one another over how to answer these questions. But what these pioneers of science all had in common was an unshakable faith in the God revealed by the Bible. They also had in common an uncanny ability to think new thoughts, to find new solutions to old problems and discover new areas for exploration. By doing so, each of them made tremendous contributions to science, often advancing human knowledge in ways that would change the world.

Not all the scientists below started their careers as Bible-believing Christians; in some cases, unbelievers became believers in the course of gathering data in their field. This trend especially shows up among the archaeologists. Of course, not all the pioneers of science were Christians. But the fifty believers we have chosen (and we could have included many more) are sufficient to tell the story of how humans acquired knowledge in each field of science. A reading of these summaries will provide a basic history of the foundation and early growth of our present scientific disciplines.

Fifty Believers Who Led the Way in Science

1. Louis Agassiz—father of glacial science (1807-1873; Swiss and American): Having studied Swiss glaciers and the phenomenon of boulder drift all over Europe, Agassiz proposed that great ice sheets—like those now in Greenland—once covered the continent. Although he is honored today especially as the discoverer of the ice ages, Agassiz should also be remembered as a zoologist and geologist whose contributions helped to establish the new science of paleontology. His extensive classification of living and fossil fish surpassed the work of all before him.

Agassiz called each species of animal or plant "a thought of God." Science historians often express consternation that a man who possessed such deep knowledge in paleontology should so strongly reject Darwinism in favor of independent successive creations, but that is where the fossil evidence led him. He saw in the similarities between living things, not evidence of evolution, but "associations of ideas in the Divine Mind." His father was the last in a long line of Huguenot pastors.

2. William Foxwell Albright—foremost archaeologist of the 20th century (1897-1971; American): To say that Albright made innumerable contributions to Middle Eastern archaeology would be an understatement. No archaeologist before or since has attained to his level of achievements across the spectrum of requisite disciplines for judging the reliability of the Bible's documents (Semitic languages, Middle Eastern history, and archaeological techniques). Though initially approaching the biblical stories as if they were fiction, Albright's discoveries turned him into a believer in the Bible's historicity. He concluded: "The excessive skepticism shown toward the Bible by important historical schools of the eighteenth and nineteenth centuries, certain phases of which still appear periodically, has been progressively discredited. Discovery after discovery has established the accuracy of innumerable details, and has brought increased recognition to the value of the Bible as a source of history."

3. Charles Babbage—creator of the computer (1792-1871; English): Renowned today as a great mathematician and inventor, his ideas were so forward-thinking that the British government could not understand the use of his inventions and refused him support. Babbage invented the first speedometer, the principle of the analytic engine, and the first true, automatic computer with information storage and retrieval, including the ability to tabulate numbers up to 20 decimals. His mathematical analysis of miracles in the Bible was the last contribution to the famous *Bridgewater Treatises,* a series of Christian apologetics writings.

4. Francis Bacon—father of the scientific method (1561-1626; English): In order to find truth, he advised two steps: first, the elimination of common prejudices (which he called idols) and second, the application of his famous "scientific method." This method of induction involves observation, hypothesis, and experimentation. A student of the Scriptures, he wrote: "No one should maintain that a man can search too far, or be too well studied in the book of God's Word or in the book of God's works; divinity or philosophy; but rather let men endeavour an endless progress or proficience in both."

5. Roger Bacon—forerunner of the scientific method (1214-1294; English): Anticipating the scientific method, his written works present the first clear case for modern, experimental science, and they point out four major obstacles in grasping truth: unworthy authority, custom, prejudice, and a pretentious show of knowledge to hide ignorance. He was the first to recognize "laws" of nature (speaking of the optical laws of reflection and refraction). His experiments with mirrors and lenses led to the invention of spectacles shortly after his death. They also contributed to the later invention of the microscope and telescope. He wrote of the sphericity of the earth and of a future in which people would travel through the air around it. His devotion to Christ led him to become a Franciscan in his forties; he regarded theology as the end of all knowledge. Because of his opposition to almost all dogmatism, however, he was in constant trouble with his religious superiors.

6. John Bartram—first American botanist (1699-1777; American): Linnaeus (q.v.) called him the "greatest living natural botanist." A self-educated naturalist, Bartram explored the American forests from Canada's Lake Ontario to Florida, classifying plants and sending information to England as official "botanist to the king for the American colonies." In 1739 he hybridized flowering plants, apparently the first American to do so. He was known as an independent-thinking Quaker.

7. Sir Charles Bell—first to extensively map the brain and nervous system (1774-1842; Scottish): one of the world's greatest anatomists, his contributions tremendously advanced our knowledge of the brain and nervous system; he discovered the different tasks of sensory and motor nerve filaments. Bell wrote *The Hand; Its Mechanism and Vital Endowments, and Evincing Design, and Illustrating the Power, Wisdom, and Goodness of God.*

8. Robert Boyle—chief founder of modern chemistry (1635-1703; English): Boyle achieved fame as the first to distinguish between a mixture and a compound, the man who turned alchemy into chemistry, and the originator of what is now known as "Boyle's Law," which states that the volume of a gas is inversely proportional to the pressure. Though his many contributions to physics (discovering the part air plays in carrying sound;

investigating specific gravities, refractive powers, electricity, crystals, etc.) would have been enough to establish his place in science, he is best known for his contributions to chemistry, many of which are presented in his first book in that field, *The Skeptical Chemist*. His critical assessment of common mystical notions turned chemistry into a true science.

In 1860 he was elected president of the Royal Society, but he declined the offer because of his Scriptural convictions against oath-taking. He learned Hebrew, Greek, and Syriac in order to study the Scriptures in the original languages. He gave a large portion of his money to Bible translation work in order to bring God's Word to those who didn't have it, and he founded the "Boyle lectures" as an apologetics series to reach unbelieving skeptics.

9. William Buckland—foremost English geologist before Charles Lyell (1784-1856; English): While professor of geology at Oxford, this committed Christian (and Anglican clergyman) became known for his systematic study of Great Britain's geologic structure. Buckland twice served as president of the Geological Society. He gave the name "alluvium" to the sediment deposited by rivers within recent times, while he reserved the name "diluvium" for the mixture of rock and bones of tropical creatures he found in English caves. His *Reliquiae Diluvianae* made the case that these deposits could be attributed to Noah's flood, though he later repudiated this as a specific evidence of that catastrophe. Many of his other findings, however, supported Cuvier's theory of multiple catastrophes.

Buckland's geological research also helped to demonstrate the evidence for an extremely old earth, as opposed to the young earth proposed by Ussher's chronology. He is also famed for his two-volume treatise, *Geology and Mineralogy Considered With Reference to Natural Theology*. Though little spoken of today, "natural theology" was a regular subject of study among these pioneers of science.

10. George Washington Carver—America's most prominent agricultural researcher and developer (1864-1943; American): Born of slave parents, Carver worked his way through school in the North, became a botanist and chemist, and turned down lucrative offers in order to devote his life to agricultural research—specifically to improve the economic conditions of the south and of his people. He showed southern farmers how to diversify their crops by planting soil-enriching peanuts and sweet potatoes instead of soil-draining cotton. Carver's research resulted in over 400 different products that could be made from these two crops, since they were otherwise overabundant as foodstuffs. He pioneered the production of synthetic marble from wood chips long before plastics were first produced from wood wastes. A devout Christian, Carver always gave God the credit for his accomplishments. At age 75, he received the Roosevelt medal, upon which was engraved: "To a scientist humbly seeking the guidance of God and a liberator to men of the white race as well as the black."

11. Georges Cuvier—founder of the studies of paleontology and comparative anatomy (1769-1832; French): His almost exhaustive work in classifying living and fossil animals resulted in many publications, most notable of which is his five-volume *Le Regne animal distribué d'aprés son organisation*. Cuvier divided the animal kingdom into four categories: vertebrate, mollusk, articulate and radiate. His theory that a series of catastrophes wiped out whole species on a number of occasions has finally been vindicated by paleontologists in recent decades. A French Lutheran and an ardent Bible-believer, Cuvier believed that the last of these catastrophes could probably be equated with the Genesis flood. Cuvier came into conflict with the evolutionary views that were already being taught in his time: he held that each animal type was specially created and that each organ served a unique purpose for that creature.

12. John Dalton—father of modern atomic theory (1766-1844; English): Like all Quakers (and other dissenters from the Church of England), young Dalton was not allowed to attend any of the English Universities, but was educated first by his father and then in the Quaker schools. He went on to teach himself mathematics and chemistry, making his own instruments to perform what others considered to be crude experiments. He also kept a meteorological diary in which he eventually entered over 200,000 atmospheric observations. His contributions to physics and chemistry included his law of gas pressures (now called Dalton's Law) and his early description of color blindness (now called Daltonism). His own color blindness sometimes caused him to inadvertently wear loud colors in violation of Quaker practice.

Dalton made one of the greatest advances in the history of science when his many observations led him to make revolutionary hypotheses about the weight and number of the ultimate particles in gases. That these ultimate particles should have differing weights was a bold departure from previous thinking, bequeathed to us by the Greeks. His deductions were brilliantly borne out by experiment. Dalton composed the first table of elements (discovering 20 of them), relating their atomic weights to hydrogen, the lightest. He correctly concluded that "all the changes we can produce [in the elements] consist in separating particles that are in a state of cohesion or combination, and joining those that were previously at a distance." Dalton practiced his Quaker faith all his life, earning a reputation as a pious man who refused to bow to any authority but God, even when British custom demanded it. In his later years, Oxford University—a school that would not accept him as a student—awarded him a doctorate.

13. René Descartes—greatest French philosopher, inventor of analytic geometry (1596-1650; French): Applying algebra to geometry, Descartes gave later scientists like Newton the ability to make the calculations that provided us with our first quantitative understanding of nature's laws. But his greatest accomplishment was to effectively lead philosophy away from medieval scholasticism and into the clear reasoning that results from questioning all past authorities. His orderly philosophical system began by doubting all things, except for the self-evident fact that he existed. Starting with the principle, "I think; therefore I am," he began a line of deductive thinking in order to compose a theory of the universe. According to his second proposition, God is the necessary and independent cause of all bodies in the universe. From these two propositions it became possible to make an orderly study of all things.

Descartes is included in this list, not because his Christianity was a model of doctrinal purity, but because his intense belief in God stands in contrast to the agnosticism that many today associate with great minds. Though a member of the established Church, he feared it when he wrote a book that agreed with the Copernican hypothesis about the earth's movement around the sun. He spent most of his last twenty years in Holland, where he enjoyed greater intellectual freedom.

Descartes' writings never take us beyond what he feels may be known by reason alone, and most have taken this to mean that Descartes himself believed in an indifferent God who merely endowed the universe with its initial, well-planned motion. Considering the self-imposed restrictions on his writings, this conclusion about his personal beliefs is not a necessary one. After all, Descartes reasoned not only that God exists, but that He must be very like the God of the Bible: omniscient, omnipotent, immaterial and separate from His creation (precluding the idea of pantheism). Even within the constraints of his published works, Descartes acknowledged the existence of the human mind as separate from matter; and he acknowledged the personal dependence of every soul on God for immortality. Learning from Descartes what may be known by reason alone whets one's appetite for what may be known by revelation.

14. Jean Henri Fabre—chief founder of modern entomology (1823-1915; French): His painstaking, direct observations of insect behavior made him the world's most knowledgeable person about insects. He imparted this knowledge in many books; his 10-volume set, *Souvenirs entomologiques* was crowned by the Institute of France. Like his friend, Louis Pasteur, he scoffed at the then current theory of spontaneous generation, the theory that life might form from non-life in a closed container. His research also led him to vigorously oppose the theory of evolution. Certain of his books were used as textbooks in French state schools, although many opposed his frequent references to divine design. Fabre spoke openly of his Christian faith. Concerning the relationship between his science and his faith in God, he wrote: "Without Him I understand nothing; without Him all is darkness . . . . You could take my skin from me more easily than my faith in God."

15. Michael Faraday—discoverer of electromagnetic induction and founder of electromagnetic field theory (1791-1867; English): Electricity had been known for several thousand years prior to his time; this humble Puritan is best known for finding out how to put it to use. Related to his discovery of electromagnetic induction (converting magnetism into electricity) are his invention of the electric motor (converting electricity into mechanical motion), the transformer (transferring electric current from one wire to another by means of magnetism), and the electric generator (converting magnetism or mechanical motion into electricity on a sustained basis). His Scriptural views on human fallibility led him to keep trying experiments where others had failed. It also led him to question all their assumptions.

When the Prime Minister, Sir Robert Peel, saw a demonstration of Faraday's generator at his lab, he asked the inventor, "What good is it?" Faraday answered: "I'm not quite sure. But I'll wager that one day your government will tax it." Many felt that Volta's battery or steam power would always be more practical than Faraday's inventions. But by the 1880s, buildings from New York to Moscow were lit with electricity produced by Faraday's generators and transported by means of his transformer.

Not content to merely find uses for electricity, Faraday sought to find out what it was. He devised ingenious experiments to prove that all forms of electricity then distinguished as separate phenomena—friction, voltaic, chemical, magnetic, thermo, and animal—are identical in nature. He also formulated the two laws of electrolysis which have ever since formed the basis of electrochemistry. With his friend William Whewell (q.v.), he introduced the terms *anode, cathode, electrode, ion, electrolyte,* and others. The electrical unit of capacity is today known as the "farad."

Because of his conviction that all creation is the handiwork of the same Grand Designer, he sought a fundamental unity between forces (as he had already succeeded in demonstrating with electricity and magnetism). While seeking a connection between electromagnetism and light, he experimented with polarized light; he found that its direction of vibration can be changed if a powerful magnet is brought close to a polarizing glass. Faraday's studies thus laid the foundations for modern theories that relate light to electromagnetism.

Further, Faraday proposed that space itself is possessed of important properties. His conception of "lines of force" laid the foundations of modern electromagnetic field theory. Faraday wrote that electromagnetic waves existed. His ideas provided the basis for Clerk Maxwell's work (q.v.) with electromagnetic waves and for the later discovery of radio waves. In this century, Einstein used Faraday's concepts in his theory of gravitation. Had Nobel prizes been awarded in his day, some say Faraday's discoveries should have brought him at least five of them. Faraday is often acknowledged as the greatest experimental genius of all time. "He smells the truth," said German professor F. W. Kohlrausch.

All these accomplishments are the more remarkable considering the fact that no British university would allow him entrance, since, as a Sandemanian, he was classed among the "dissenters" from the Church of England. Sandemanians were characterized by their dependence upon the Bible instead of church tradition. Further, his extreme poverty prevented him from receiving training in any of the dissenting schools. All his knowledge was gained by his own reading and experiment. In 1832 Oxford University awarded him an honorary doctorate. By the end of his life, Faraday had received almost a hundred medals or diplomas from various institutions. He refused the presidency of the Royal Society and the Royal Institution, preferring to devote himself to research.

This devotion also caused him to pass by many opportunities to become wealthy in industry as a consultant. A desire to find the laws of God clearly motivated Faraday; neither money nor fame could keep him from this goal, though he made it a point to keep his personal beliefs out of his public lectures on science. However, when Faraday lay on his deathbed, an interviewer asked him about his speculations on the hereafter. Faraday answered: "Speculations? I have none. I am resting on certainties." And then he quoted 2 Timothy 1:12: "I *know* Whom I have believed, and am persuaded that He is able to keep that which I have committed unto Him until that day."

16. John Flamsteed—maker of the first modern star catalogue (1646-1719; English): Because the observations he made from his Greenwich observatory became the model for all modern star catalogues, today's global maps continue to use Greenwich, England as their prime meridian. As the first astronomer royal, Flamsteed's observations were made with limited means; he supplied all of his instruments out of his own salary. Yet, he took full advantage of his position, making extremely accurate observations with the telescopes available in his day. Because he was a perfectionist, he came into some conflict with Isaac Newton and Edmund Halley, who desperately wanted his observations before he was ready to publish them. Flamsteed's three-volume *Historia coelestis Britannica* contains his observations and his catalogue of nearly 3,000 stars. Flamsteed was also an ordained clergyman, and is said to have been devout in his daily living and Christian ministries.

17. John Ambrose Fleming—inventor of the diode (1849-1945; English): Because of his contributions to electronics, wireless telegraphy, and television, some call him father of modern electronics. In 1904 he invented the vacuum diode tube, first called "the Fleming valve." This evacuated glass bulb contains two electrodes, one that emits electrons and the other that accepts electrons, so that electricity travels in just one direction. For half a century, the diode became an indispensable part of all radio receivers and transmitters, as well as television sets, computers, etc. (until the invention of the transistor). Edison and Marconi consulted Fleming in order to complete their inventions. The son of a Congregational pastor, Fleming is also known for his work in Christian apologetics. He was particularly alarmed at the growing acceptance of evolution as fact. Thus he formed the "Evolution Protest Movement" and wrote a book defending the biblical account of creation.

18. Nehemiah Grew—co-founder of the science of plant anatomy (1641-1712; English): Both as a plant anatomist and physiologist, he made great contributions to science. Grew is credited with the discovery that plant stamens are male organs, and he developed the first theory to account for pollen. All studies of comparative organs stem from Grew's *Comparative Anatomy of Stomachs and Guts Begun,* in which he first used the word "comparative" in this way. He was also the first to describe the precisely individual make-up of finger ridges, a knowledge which was much later put to use in fingerprinting.

But Grew's greatest accomplishments came about as the result of his humble deference to Italian botanist, Marcello Malpighi. The history of science is better known for its frequent rivalries and intense disputes. In this case, after seven years of intense investigations into plant anatomy, Grew learned that Malpighi's own researches had duplicated much of his own work. At a meeting of the Royal Society of London in 1671, Grew praised the Italian's work and offered to retire from further studies in the field. The Royal Society instead encouraged Grew to continue his research, and Malpighi was so impressed with Grew's actions that he began a cooperative working relationship with him. Together, the pair laid the foundation for plant anatomy that was not improved upon for a hundred years.

The son of an anti-Royalist clergyman, Grew entered his investigations with little scientific training but with a firm conviction that "both plants and animals came at first out of the same hand, and were therefore the contrivances of the same wisdom." But even his Christian upbringing did not prepare him for the perfection and purposeful order he observed in his research. He wrote that even the plainest walking stick "far surpasses the most elaborate woof or needle-work in the world."

19. Stephen Hales—first to bring the exacting methods of physics to biology (1677-1761; English): This clergyman ranks with the greatest of physiologists, chemists, and inventors. Hales is remembered for introducing rigorous quantitative methods to the study of animal and plant physiology. He made the first quantitative measure of blood pressure; his experiments led to the development of the instruments still used for measuring blood pressure today. Among his inventions were devices to distill fresh water from sea water and ventilators for pumping fresh air into ship holds, hospitals, and jails.

Hales carefully measured the water consumption of plant leaves and roots; but his deductions went far beyond what these calculations alone could tell him. He was the first to claim that plants absorbed air through their leaves, converting it into solid substances. He also made the first deductions about the chemistry of the air and about the fact that leaves actually processed light for the plant's use. Such inspired hypotheses, made many years ahead of their time, can in part be attributed to his expectation of perfectly designed laws by which God governed the universe. As Hales wrote: "The farther researches we make into this admirable scene of things, the more beauty and harmony we see in them: And the stronger and clearer convictions they give us, of the being, power and wisdom of the divine Architect."

20. Joseph Henry—discoverer of the principle of self-induction (1797-1878; American): The henry, the standard unit of electrical inductive resistance, is named for him. He shares credit with Michael Faraday (q.v.) for discovering electromagnetic induction. Many of his experiments resulted in discoveries that were later built upon by others. Henry invented the earliest version of the electromagnetic motor; he also invented a powerful short-coil magnet, which is essentially the same as that now used today in electric motors and generators. Further, he discovered the laws that led to the invention of the electric transformer.

Using his long-coil intensity magnet, he first demonstrated long-distance transmission of electrical current, paving the way for the commercial telegraph. While serving as the first director of the Smithsonian Institution, he organized a corps of weather observers and supervised them for 30 years. Their successful work led to the creation of the U.S. Weather Bureau. A committed Christian, Henry was also known for his dependence upon God in all his work: standard operating procedure in all his experiments included prayer for guidance in every major decision.

21. Sir William Herschel—discoverer of Uranus (1738-1822;

English): After serving with the Hanoverian Guards in the Seven Years' War and working as a professional musician, he turned to astronomy with such a passion that the telescopes of his day could not satisfy his many curiosities about the heavens. Thus he built the largest telescopes of his time, discovering two new moons for Saturn and that Mars had white poles. Herschel was also the first to identify binary star systems (stars that orbit each other). His greatest discoveries were made as a result of his systematic study of the patterns and distribution of the stars. Called the founder of sidereal astronomy for his exacting methods, he is most famed for his discovery of a new planet outside the orbit of Saturn (later named Uranus). This discovery brought him up from relative obscurity to a position as "the king's astronomer."

Herschel is also renowned for his conclusion that our solar system is situated within and near the edge of the Milky Way, which he claimed was shaped somewhat like a thick pancake. Although the scientific consensus would not catch up with him for another century, Herschel maintained that all the stars we could discern belonged to the Milky Way, while other galaxies like ours existed far away. His Christian devotion was demonstrated not only by his church associations (playing the organ for church services) but by his many comments on God's sublime harmonies in the heavens. "The undevout astronomer must be mad," Herschel said.

22. Sir William Huggins—first to measure the stars' velocities and chemical composition (1824-1910; English): Using spectroscopy, he was the first to break down starlight into its constituent colors in order to learn the stars' chemical composition. He discovered that stars were composed mainly of hydrogen and contained much smaller amounts of other elements. Huggins also showed that the "unresolved" nebulae (those that are not classified as galaxies today) could not be resolved from their spectra into stars because they were composed of gas, not stars. As far as cosmologists are concerned, his greatest accomplishment may have been his recognition of blue- and red-shifting starlight as a Doppler effect. It was this discovery that later led to the further discovery that our universe is expanding. Huggins readily confessed his faith in Christ.

23. James Joule—discoverer of the first law of thermodynamics (1818-1889; English): Joule attacked the caloric theory of heat, the view that heat was a fluid that combined chemically with particles of matter. By demonstrating that a specific amount of work produced a specific amount of heat, and vice versa, he replaced the caloric theory with the kinetic theory, attributing heat to the motion of the molecules themselves.

Joule's experiments demonstrated the relation between mechanical, electrical, and chemical effects; this led to his principle of the interchangeability of various kinds of energy and to the principle of energy conservation. Also called the first law of thermodynamics, this principle states that matter and energy can be neither created nor destroyed, but each may be converted into the other. As a committed Christian, Joule thus appears as an archetype of the many scientist-believers who first recognized patterns that make up our most fundamental laws of science. Using strict, quantitative methods in a long chain of experiments, he determined the constant known as "the mechanical equivalent of heat": the amount of work required to produce a unit of heat. Today energy units are measured in "joules."

24. Kelvin, William Thomson—first to clearly state the second law of thermodynamics (1824-1907; British): Deeply influenced by James Joule, Lord Kelvin was one of the first scientists to adopt the concept of "energy." As a result of his establishment of an absolute scale of temperature, intervals of temperature are now measured in degrees which bear his name. During his prodigious career, he published over 300 papers, touching on nearly every area of physical science. In one he redefined Joule's conservation of energy and went on to state the principle of energy dissipation, now known as the second law of thermodynamics.

Kelvin is also known for his many nautical inventions—such as ship's compasses and devices for ships to take soundings—as well as for solving the problems of sending telegraphic signals over long distances through the first transatlantic cable. Kelvin's biblical faith led him to strongly support the Bible's teachings in British schools. He also opposed the doctrines of Darwinian evolution. Evolution, as he saw it, demanded an infinitude of time, whereas thermodynamic considerations rightly led him to believe that the universe and the earth have a limited history.

25. Johannes Kepler—discoverer of the laws of planetary motion (1571-1630; German): Kepler led science to make a complete break with the Ptolemaic tradition that had held sway for 2,000 years. The earth, he maintained, was a planet that, with the others, orbited the sun. Copernicus had said it; Kepler proved it. But Copernicus had believed the planets to move in circular paths; he had also perpetuated the tremendously complicated system of epicycles in order to explain the erratic paths and velocities of the planets. Trained for the ministry, Kepler felt that a believer in a truly orderly creation should look for a simpler explanation. And so he persevered through thirty years of illness, many severe misfortunes, and persecution for his Protestant religious beliefs, seeking the simplicity and order that he could at first know only by faith. Kepler said that he was merely "thinking God's thoughts after Him."

Not long after Kepler had been expelled from Graz (along with all Protestant theologians), he joined the great mathematician Tycho Brahe in

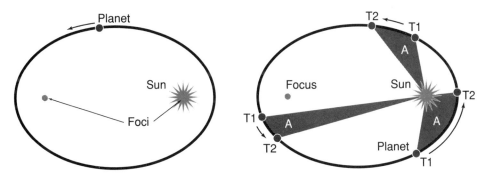

Kepler's first and second law. According to Kepler's first law, the orbit of every planet forms an ellipse, with the sun as one of the two foci. According to his second law, a planet's radius vectors sweep through equal areas (A) in equal times. *T1, T2*, etc. indicate the positions of the planet as it passes through equal intervals of time. —*Rick Vogeney*

Prague. Adding his own heavenly observations to Brahe's records, Kepler painstakingly made calculations in order to eliminate every conceivable, orderly explanation for the planetary orbits until he eventually arrived at one that worked. Brilliantly combining deduction with observational, inductive methods, Kepler showed that the planets move in ellipses, not circular paths, with the sun at one focus. In his further search for regularity, he discovered that the planets' radius vectors sweep through equal areas in equal times. Thus he precisely accounted for the irregular velocities of the planets and eliminated the need for complicated epicycles. Going still further, he proposed a correlation between the periods and distances of the planets. And so Kepler established his three laws of planetary motion that brought order to the chaos of astronomy—and some say, to science as a whole.

Suggesting that some kind of force from the sun acted on the planets, Kepler was the first to try to propose a physical explanation for the motions of the solar system. Kepler's laws and theories laid a solid foundation for Newton. Though he was forced to make his living by drawing up astrological charts, Kepler is more responsible than anyone for separating astrology from astronomy. However, his efforts to find a correlation between his planetary laws and the intervals in the musical scales have more in common with the mystical arts. But if Newton can be forgiven for pursuing alchemy, Kepler can be excused for practicing astrology, the vice of his field. To the extent that Kepler sought an orderly creation, as described in Scripture, his work was foundational to modern science. To the extent that he sought a mystical music of the spheres, as taught by the Greeks, he wasted his great efforts on fruitless by-ways.

26. John Kidd—pioneer in the development of chemical synthetics (1775-1851; English): This Oxford don's work to extract chemicals from coal led to the compound building processes from which all modern synthetics are derived. Kidd pursued this ground-breaking work in part to demonstrate that God had prepared the natural world for man's use. Known for his Christian convictions, he used his platform as a naturalist to tell others of the evidence of divine wisdom seen in nature. Kidd wrote one of the famous Bridgewater Treatises, *The Adaptation of Nature to the Physical Condition of Man,* to defend biblical faith against the rising tide of naturalism.

27. Gottfried Wilhelm Leibnitz—co-inventor of calculus (1646-1716; German): His greatest single accomplishment may have been his independent development of calculus (independent of Isaac Newton, q.v., who was given priority, much to Leibnitz's dismay). But to allow Leibnitz to be remembered merely for this is to miss his substantial work in fields outside of mathematics: philosophy, theology, geology, and even political diplomacy. First among these must be his contribution to philosophy. Leibnitz expounded a logically-connected system that reconciled the systems of many predecessors, but he is most renowned for his distinct demonstration of God's existence.

Rather than arguing from design (the teleological argument) or from the idea of God (the ontological argument), or from the moral argument (Kant's favorite), Leibnitz chose his own special form of the causation (cosmological) argument: "the principle of sufficient reason." By this he meant that "for everything there must be a sufficient reason why it is so and not otherwise." Leibnitz showed that there is no sufficient reason for the universe's existence apart from the choice of God. Because a God of perfection wills the best, this universe must be "the best of all possible worlds," a concept made most famous by Voltaire's later criticism of it.

But Voltaire's mockery of this idea in *Candide* is convincing to a philosopher only insofar as he is willing to side with the hedonistic view. This cannot be the best of all possible worlds if one associates "good" and "value" merely with pleasure and the absence of pain, rather than with the development of spiritual maturity or moral perfection. Among Leibnitz's other well-known works are his defense of the doctrine of the Trinity and his contribution to the development of probability in logic.

Leibnitz founded and served as first president of the Berlin Academy of Sciences. In *Protogaea,* his work on the origin of the earth, he proposed that our earth's center is molten. Reasoning from the simple observation that the exposed rock of our earth's crust is composed mainly of "the imperishable silicates . . . almost entirely vitrified," Leibnitz proposed that our entire earth must have cooled from a once-molten condition.

Leibnitz is also known—perhaps as no other person in history—for his efforts to re-unite Christians. In his travels and writings, he repeatedly exhorted those who called themselves Christians to stop wasting their energies against one another. Leibnitz expended considerable efforts in various schemes to persuade the "Christian" nations to stop fighting one another, Catholics to re-unite with Protestants, and Lutheran churches to re-unite with Reformed churches.

28. Carolus Linnaeus—father of taxonomy (1707-1778; Swedish): The universal system of classification of plants and animals is named after him. Linnaeus laid down the principles and nomenclature we still use today to define genera and species. Using this binomial nomenclature, modern scientists continue to identify each species by two words: first, the genus to which it belongs, and second, a word of further description. Other botanists followed his example, making it possible for the first time to easily trace an organism from one author's work to another. Linnaeus carefully arranged 6,000 species into genera, establishing the names still used today.

His belief in the Scriptures led Linnaeus to seek evidence for both creation and flood.* The Linnaean system was inspired by his search for the distinct "kinds" of created organisms related in Genesis. His early belief in the "fixity of species" was based on his observations of a limited group of species in Scandinavia. Later, when Linnaeus had been exposed to a much wider range of specimens, the evidence convinced him that species do change and that species must originate from the broader category he called genera. His decision to reject the notion of common descent for all living things was based not only upon Genesis, but upon the evidence of distinct types and the lack of transitional forms. Contrary to the impression given by many modern accounts, the taxa are thus not "evolutionary classifications" at all. Darwin merely came up with a theory to try to explain the existence of the distinct categories that had already been recognized by a devout creationist.

29. Joseph Lister—founder of antiseptic surgery (1827-1912; English): In a day when untold numbers died after surgery from "hospital fever," "hospital gangrene," and other mysterious problems, Lister sought the cause and the cure; he changed forever the standard procedures for surgery, preventing countless deaths. Putrefaction in open wounds, according to the consensus, was caused either by oxygen in the air or by spontaneous generation of germs, meaning that it was a necessary evil of surgery. Learning from Pasteur's experiments (q.v.) that putrefaction might be due to microbes coming from the air, Lister strove to find ways to eliminate the microbes. His

*Both his biblical interpretations and his observations of nature led Linnaeus to maintain that the flood had been a relatively tranquil event, not one that reshaped the earth.

extensive experiments led to his use of a careful mixture of chemical agents to serve as disinfectants, sulphochromic catgut to serve as sutures, and the scrupulous cleansing of operating instruments and personnel.

In addition to the development of such standard procedures, Lister invented many new instruments, including the aortic tourniquet, the sinus forceps, and the wire needle. Lister founded the British Institute of Preventive Medicine, later named the Lister Institute. Throughout his life, and in spite of his many honors, Lister maintained the humility characteristic of his Quaker upbringing, as well as the biblical beliefs handed down to him. At a time when many scientists of his acquaintance denied divine revelation, Lister continued to describe himself as "a believer in the fundamental doctrines of Christianity."

30. James Clerk Maxwell—formulator of the electromagnetic theory of light (1831-1870; British): After reading Michael Faraday's *Experimental Researches,* Maxwell turned his well-trained mind to the study of electricity, eventually devoting himself to building a mathematical framework for Faraday's electromagnetic laws. Like Faraday (q.v.), Maxwell rejected the notion that magnetism, electricity, or light could produce their effects at a distance without the space between having some properties to make this possible; and so he sought explanations involving a medium or field. Light, he theorized, consists of waves or vibrations in this field. In fact, he said that electricity and magnetism must be different forms of the same phenomenon. To support his electromagnetic theory of light, he confirmed experiments to show that light in space and electricity in a wire traveled at about the same velocity. Maxwell proposed that all these phenomena lay at different sections of the electromagnetic wave spectrum. His achievement laid the necessary foundations for the discovery of radio waves.

Maxwell is also famous for his precise, mathematical accounting of the properties of gases: their heat conduction, diffusion, viscosity, etc. Maxwell's law of gases is still used in calculations today. Known as a diligent student of the Bible and a man who freely confessed Christ as his Savior, he wrote that God's command to subdue the earth provided the ultimate motivation for his scientific research.

31. Gregor Mendel—father of genetics (1822-1884; Austrian): Like Roger Bacon long before him, Mendel served his God both as a monk and as a devoted student of His creation. Mendel discovered his principle of heredity and laid the foundation for the science of genetics in a small monastery garden. Before his experiments with garden peas, no one had made such careful observations over the required generations in order to determine the statistical results of cross-breeding. And even after his work had appeared in principle libraries in Europe and America, another generation passed before anyone recognized his principle of heredity as the great key to all the biological sciences. Patient observations led him to conclude that the characteristics of living things are determined by inherited, paired units, now known as genes.

We might well wonder if Darwin would have extended his observations of microevolution (small changes) into macroevolution (the view that all species are descended from a common ancestor) had he been aware of Mendel's findings. Darwin held the commonsense (but erroneous) notion that offspring receive traits that are intermediate blends between their parents. This allowed him to postulate that evolution was unlimited, since he thought that the original traits were lost forever.· Had he understood Mendel's finding that offspring actually receive a dominant (observed) and recessive (hidden) gene for each trait, he may have decided, as had others, that natural selection actually acts as a conservative force, not as a force that would promote unlimited change, forever leaving behind old traits.

32. John Michell—father of seismology; predictor of black holes (1724-1793; English): This pastor/geologist is known as the creator of the science of seismology. He suggested that earthquakes sent out their shocks from central points in the form of waves; the source of a quake could thus be determined by learning the precise times the waves reached various locations. By inventing the torsion balance, he made it possible to measure forces too minute to detect by other means.

In addition to making sensitive measurements of the activity beneath the earth's surface, Michell set up the first laboratory experiment to accurately measure the force of gravitation. By suspending a large and a small lead sphere, he set up his torsion balance to measure the actual gravitational attraction of one laboratory object upon another. Although the design and the apparatus itself was his, he died before he was able to use it, bequeathing the experiment and the apparatus to Henry Cavendish, who used it to make the first measurement of G.

Astrophysicists also honor Michell today as the first to conceive of black holes. He wrote about them to Cavendish in great detail nearly two centuries before American physicist John Wheeler coined the term "black hole" in 1968. Understanding escape velocities, Michell correctly deduced that light itself, if composed of corpuscles (today called "photons"), would not escape from an object of sufficient mass. He calculated that any star with the same density as the sun but with a radius 250 times larger would have an escape velocity that would exceed the velocity of light. Such a massive object would not be visible to us. Michell even determined the main method used today to detect a black hole: Any black hole that orbited another star should change the orbit of its companion, thus disclosing its presence.

33. Samuel Morse—inventor of the telegraph (1791-1872; American): Morse first gained acclaim as an artist. Although his paintings of democracy in action in Washington gained widespread recognition, Morse's income was irregular and he was forced to spend his time painting portraits rather than the historical paintings he loved. While returning from art study in Europe in 1832, he learned of the discovery of the electromagnet. Morse immediately thought of the possibility of using electromagnetism to transmit messages over long distances. Faster means of communication, he knew, would have prevented a major battle after a peace treaty had been signed at the end of the War of 1812. Faster communication would have also allowed him to learn of his wife's sudden illness while he was traveling a few hundred miles away from her in 1825; as it was, the news reached him too late for him to even attend her funeral.

Others also thought of the basic idea of an electric telegraph, but Morse developed the first working model in 1835; and only Morse persevered through the years, long after his financial partners had given up on the idea. For eleven years, he continued to produce improved models and to try to sell the idea to potential financial backers on both sides of the Atlantic, without success. He developed the "Morse code," an alphabet composed of dots and dashes; a form of which is still the universal code in use today. A public demonstration of his telegraph line across New York Harbor failed when a ship severed the line. Morse lived in poverty and often in hunger, but his perseverance can best be explained by his faith in a God of purpose. During this period, he wrote: "I am perfectly satisfied that, mysterious as it may seem to me, it has all been ordered in view of my Heavenly Father's guiding hand."

Finally in 1843 he persuaded the U.S. Congress to fund the building of a telegraph line between Baltimore and Washington, DC. When he completed the project on time and on budget on May 24, 1844, he chose as the first message, "What hath God wrought!" from Numbers 23:23. Others immediately claimed credit for his invention, but eventually, the Supreme Court decided that Samuel Morse should own sole rights as inventor. Morse is also known for bringing the process of photography to America and, while working with Samuel Draper, for improving the process so that only a one-minute exposure was required rather than a ten-minute exposure time.

Morse's inventions also penetrated the spiritual realm. He was among the first to conceive of and implement the concept of a Sunday school for children, and he promoted the idea during his travels. Though poor throughout much of his life, his wealth in his later years gave him the opportunity to be a philanthropist, supporting missionaries and schools for pastors. Late in life, Morse wrote: "The nearer I approach to the end of my pilgrimage, the clearer is the evidence of the divine origin of the Bible, the grandeur and sublimity of God's remedy for fallen man are more appreciated, and the future is illumined with hope and joy."

34. Isaac Newton—discoverer of the universal law of gravitation (1642-1727; English): This universally recognized genius successfully synthesized the work of great achievers like Descartes, Brahe, Kepler, and Galileo, providing a mathematical framework for calculating the movements of all bodies in the universe. Newton wrote: "If I have seen further it is by standing on the shoulders of giants." Still, his own astute principles were not improved upon for 200 years, until Einstein's development of the general theory of relativity.

Newton laid the foundations for his greatest accomplishments at the age of 23, when his education at Cambridge was interrupted by the Great Plague. While the school was closed, he used his extended vacation at his mother's country home to develop the binomial theorem and a whole new branch of mathematics: calculus, an innovation that has been called the greatest accomplishment in mathematics since ancient times. It was also during this period that he speculated upon the relationship between falling apples and moving planets, though his finished work on this appeared 20 years later.

Isaac Newton (1642-1727) performs one of his early optical experiments. —*Courtesy Bausch & Lomb Optical Company & AIP*

Young Newton first became known to the scientific community for his invention of the reflecting telescope (using a mirror to gather light rather than the lens of the refracting telescope), the type used by today's largest optical telescopes. Newton sent a model to the Royal Society of London, telling them he would also like to send them information on a new discovery about light.

By experimenting with sunlight passing through various configurations of prisms, Newton showed not only that white light could be separated into its component colors, each refracted to a different degree to produce a spectrum, but that these could be re-combined to form white light again. Thus he demonstrated white light to be a mixture and not a separate color like the others. Further, he showed that objects have color only because they "reflect one sort of light in greater plenty than another." He accomplished this by changing the colors of objects in a darkened room with colored lights, noting that an object's color was most vivid when illumined by the light of its own daylight color.

Newton is best known for the findings he proclaimed in 1687 in his *Philosophiae Naturalis Principia Mathematica,* a work that built upon and combined Galileo's law of falling bodies with Kepler's law of the periodic times of the planets. From this foundation, Newton developed his three laws of motion and explained in mathematical terms how the orbit of the moon is due to that body's continual fall towards the earth away from a straight line. Newton remarked on how centripetal force (a force that draws toward the center) must be precisely balanced by centrifugal force (a force that causes a stone to fly from a rotating sling) in order to produce the observed orbits. As a proof that the same force acts on both the moon and nearer falling objects, Newton correctly calculated the rate of falling bodies at the earth's surface (16 feet from rest in one second) by using the known distance to the moon and its rate of being pulled back into circular orbit.

His calculating tool was his famous inverse square law: Each object in the universe behaves as if it attracts every other object with a force proportional to the product of the masses and inversely proportional to the square of the distance between them. Thus the gravitational influence of one body upon another becomes one fourth as strong when the distance between them is doubled, one ninth as strong when the distance is tripled, and so on. From this tremendously useful discovery, others were soon able to precisely calculate the motions of planets, earth's tides, and even the return of comets. To his credit, Newton never tried to explain exactly what gravity is, but in the cautious spirit of empirical science, he merely offered a useful description of what he observed. Today, the cause of universal attraction has yet to be explained (though general relativity comes somewhat closer to an explanation).

At the end of his *Principia,* Newton stated that "this most beautiful System of the Sun, Planets and Comets could only proceed from the counsel and dominion of an intelligent and powerful Being." Though the deists later took Newton's laws as evidence for a universe that ran by itself, Newton himself believed in a God who "governs all things" and "is more able by His Will to move Bodies . . . and thereby to form and reform the Parts of the Universe, than we are by our Will to move the Parts of our own Bodies." Newton often confessed that his interest in theology far surpassed his interest in science; he spent much of his time studying the Scriptures in depth, producing books on chronologies in the Old Testament and on the book of Revelation. After many years of diligent Bible study, Newton concluded: "We account the Scriptures of God to be the most sublime philosophy. I find more sure marks of authenticity in the Bible than in any profane history whatsoever."

35. Ambrose Paré—first modern surgeon (1510-1590; French): His innovations in surgical procedures earned him a reputation as the greatest surgeon of his day. Medical historian F. H. Garrison ranks him among the three greatest surgeons in ancient or modern times (the others being Joseph Lister (q.v.) and John Hunter). Paré's achievements cannot be attributed to scholarly training, for he had none. Credit must go to his own researches and his experience as surgeon on the battlefield, along with his independent spirit. Although he had carefully studied recent and ancient traditions regarding the treatment of wounds, he rejected their painful or harsh procedures whenever he could.

At first, Paré received severe criticism from the medical community for his refusal to pour boiling oil into wounds after amputations in order to cauterize them. He was not the first to tie veins and arteries to stop their bleeding, but he perfected and standardized the procedures so that others eventually accepted them as superior to the "boiling oil technique." Surgeons of the day also routinely applied boiling oil to gunshot wounds, believing this necessary to deal with their poison. Paré developed a mild ointment to treat gunshot wounds and powder burns instead, and soon proved the superiority of his new procedures. Paré's innovations were also felt in the field of obstetrics, where he showed that many infants' lives could be saved in difficult deliveries by the operation of turning them (podalic version). Standard procedure of the day was to tear apart the infant and extract it.

Paré's reputation soon exceeded all other surgeons of his day; he became surgeon to three kings. Such success, combined with such independence of mind often produces an arrogant spirit, but Paré was known for his humility. When asked for help, he replied that he "would use all the little knowledge which it has pleased God to give me."

As a participant in the French Reformation, Paré was a student of the Scriptures and subscribed to the biblical doctrine of "justification by faith." His life was spared at a time when thousands of Protestants in France were slain. Some say he escaped death during the Massacre of St. Bartholomew's Day in 1572 only because his patient, Charles IX, hid him in his bedchamber. On another occasion, his life was spared after being captured by the Spanish because of his own compassion on wounded Spanish soldiers. Paré was known not only for his physical treatments, but for his prayers for his patients. He never took full credit for success; his oft-quoted remark is inscribed on his statue in Paris: "I treated him, God cured him."

36. Blaise Pascal—founder of probability studies and hydrostatics (1623-1662; French): Pascal is known for his contributions to physics, mathematics, philosophy, and Christian apologetics. Before he reached his teenage years, he taught himself geometry. And before the age of 20, he invented the first calculating machine. He also developed the concept of applying mathematics to statistics in such a way as to make it possible to mathematically analyze probabilities. Such analysis is now of tremendous importance to insurance companies, for biological statistics, and for all types of statistical analysis.

Pascal's writings are permeated with a true skepticism that doubted not only the value of dead formalism in religion but the reliability of scientific tradition as well. According to the scientific consensus of his day, "nature abhors a vacuum." The recently invented barometer seemed to uphold this opinion, since the mercury was supposed to be held up by nature's abhorrence of the vacuum. Pascal, however, held that nature lacked the ability to abhor anything, being merely God's creation and not God or spirit itself.

Using one of the first barometers inside a vacuum tube (a vacuum inside a vacuum), Pascal set up an experiment where the height of the mercury could be measured against another tube at various altitudes. He wrote:

> If it happens that the height of the quicksilver is less at the top than at the base of the mountain . . ., it follows of necessity that the weight and pressure of the air is the sole cause of this suspension of the quicksilver, and not the abhorrence of a vacuum . . . one cannot well say that nature abhors a vacuum more at the foot of the mountain than at its summit.

Pascal thus disproved the notion that nature abhors a vacuum, while also demonstrating that air has weight or pressure. Air, he said, is like a great bulk of wool many fathoms high, compressed by its own weight so that the bottom layers are most compressed. The laws of pressures Pascal developed in fluids led eventually to such advances as the modern car's hydraulic brakes.

In a sense, Pascal extended his theory of probability to life's greatest issues. If reason cannot "prove" God's existence beyond all doubt, then every person must make his wager. Pascal's famous wager is a challenge to skeptics of all ages; to bet on God, he says, is a win-win proposition. If God turns out not to be there, the Christian has lost nothing (and may in fact live a more fulfilling life than nonbelievers); if God *is* there at the end of one's life, then the Christian has gained everything while the unbeliever has lost everything—for all eternity. "Let us assess the two cases," he said. "If you win, you win everything. If you lose, you lose nothing."

Dissatisfied with nominal Christianity, Pascal taught that each person must have a personal relationship with Christ; only then can God empower us to truly love and serve one another. Pascal disliked Descartes' purely rational philosophizing about God. Ironically, some say it was Descartes' compassionate visit to the ill Pascal that prompted Pascal to write: "The heart has reasons which reason knows nothing of."

37. Louis Pasteur—formulator of the germ theory of disease (1822-1895; French): Pasteur laid the groundwork for all his later discoveries by studying the process of fermentation. He showed that abnormal fermentation of wines and beers could be prevented by heating them to about 135° F for several minutes, a treatment that led to the standard practice of "pasteurization" of dairy products as well. After proving that fermentation was due to the presence of a microorganism (yeast), he eventually generalized this germ theory of fermentation to a germ theory of disease.

To demonstrate this, Pasteur first set out to disprove "spontaneous generation," the popular notion that living organisms can arise from non-living matter or from dead organic matter. Earlier scientists had even given directions for the creation of mice from old rags or flies from rancid meat. Pasteur's new explanations were especially pitted against those of French biologist Felix Pouchet and later by English pathologist Charles Bastian, both of whom claimed that microorganisms could be generated merely by bringing together decaying organic matter, air, and water.

When Pasteur first heated these substances to kill the already existing bacteria, his opponents claimed that he had merely altered the chemical composition, making spontaneous generation impossible. Pasteur carried out an ingenious series of experiments in which he continued to heat the components but admitted specially filtered air (or pure air from the Alps). Through these experiments he eventually answered all his opponents' objections and demonstrated that the microorganisms in his opponents experiments were not spontaneously generated, but had come from the air.

As a man with deep-seated Christian convictions, Pasteur also disliked the concept of spontaneous generation because it reduced creation to a mere chemical reaction. Like the concept of biological evolution, this notion violated both his scientific sensibilities and his biblical understanding. Pasteur continued to oppose the scientific establishment on the issue of evolution for the rest of his life.

The real proof of his germ theory of disease came through his tireless efforts to develop specific preventive measures for various diseases: silkworm disease, chicken cholera, anthrax, and rabies. Other scientists held that most of these diseases were poisons that developed spontaneously in the organism. Pasteur's theory of parasitic invasion suggested certain defensive strategies. As we have seen, Joseph Lister successfully applied Pasteur's theories to surgery, boiling his instruments in water and applying antiseptic agents to the wounds in order to prevent the growth of microorganisms.

Pasteur searched for antidotes. An accident led him to the development of the first vaccine for cholera. As Pasteur himself wrote, "Chance favors only the prepared mind." Upon returning to his laboratory after several months away, he found some old chicken broth cultures that he had left there months before. Since these had been infected with cholera, he decided to try to inject these old cultures into healthy hens. Finding that these hens did not develop cholera, he decided to inject the hens again, this time with newly infected cultures. Surprisingly, the hens still did not develop the disease. Pasteur immediately knew that by injecting the hens with a milder form of the disease, he had stumbled upon an antidote.

In the years of work that followed, Pasteur solved the many problems involved in developing vaccines for anthrax and rabies. Perhaps no other scientist has made greater contributions to the saving of human life. A devoted Catholic, he was not a man whose religion was a mere outward show. When asked about how his religious beliefs had been affected by his science, he answered: "The more I know, the more does my faith approach that of the Breton peasant."

38. William Prout—first to relate atomic weights to hydrogen (1785-1850; English): In 1815, this chemist was the first to propose that the atomic weights of the elements form a series of whole numbers, the result of being exact multiples of the atomic weight of hydrogen. Biologists in the field of digestion know him better for his 1823 discovery of the existence of free hydrochloric acid in the stomach. He demonstrated that this acid played a necessary part in gastric digestion, but that other agents were also required. A pioneer in the science of nutrition, he was the first to divide the fundamental foodstuffs into proteins, fats, and carbohydrates. Prout wrote one of the Bridgewater Treatises, a famous Christian apologetic series. In it, he reacted to the unscientific spirit of romanticism that had begun to besmirch the natural philosophy of his day (through science popularizers like Sir Humphrey Davy and T. H. Huxley):*

> The poor untutored savage "sees God in every cloud, and hears Him in the wind." The complacent philosopher smiles at the credulity of the savage, and perhaps deifies "the laws of nature!" Both are alike ignorant; nor is the imagined Supreme Being of the untaught savage in any degree more absurd than the imagined Pantheism of the philosopher.

---

* The many science popularizers who put their own pantheistic spin on science today (by speaking of how the laws of nature "govern" the universe) continue to infect science with a philosophy foreign to it.

39. William Mitchell Ramsay—foremost archaeologist of Asia Minor (1851-1939; British): Ramsay had been schooled in the Tübingen theory, which declared that the New Testament documents were written by second-century Christians who were trying to insert fabricated stories into first-century settings. This supposedly resulted in many geographical and historical inaccuracies. Though thoroughly persuaded by this and other views of higher criticism in his university studies, Ramsay was surprised to find corroboration of the book of Acts once he got on the field and examined the topography and antiquities of Asia Minor for himself. The places, the titles, and the technical terms Luke used precisely fit first century Asia Minor (see Volume 4 for details). Using this field study to scrutinize the travels of Paul in the book of Acts, Ramsay eventually reversed his position. He wrote:

> In fact, beginning with a fixed idea that the work was essentially a second century composition, and never relying on its evidence as trustworthy for first century conditions, I gradually came to find it a useful ally in some obscure and difficult investigations.

Ramsay concluded that "Luke's history is unsurpassed in respect of its trustworthiness." Because of his many discoveries, Ramsay is generally regarded as one of the greatest archaeologists of all time. His personal views in theology underwent a change from the extremes of radical criticism to acceptance of the Bible's history and doctrines, as can be seen from the more than twenty books he wrote to support and interpret the Bible.

40. Sir William Ramsay—discoverer of the rare gaseous elements (1852-1916; British): Those who knew Ramsay personally knew him to be an outspoken, evangelical Christian who had once intended to enter the ministry. Today scientists know him as the only person ever to discover a whole family of new elements. Aware that atmospheric nitrogen was heavier than chemical nitrogen, Ramsay predicted the existence of an unknown heavy gas in the atmosphere. Others attributed the difference in densities to different forms of nitrogen, analogous to the differences between oxygen and ozone. Contriving ways to remove oxygen and nitrogen from the air, Ramsay found the gas now called argon, which composes about one percent of the atmosphere.

Lockyer and Frankland had discovered the spectrum for helium in the sun in 1868, but helium itself had never been found on earth. Ramsay first obtained it on earth by heating the mineral cleveite in acid. Knowing that the position of helium and argon in the periodic table suggested the existence of at least three more inert gases, Ramsay returned to his technique of removing oxygen and nitrogen from the air to search for these yet-more-rare elements. In 1898 he found neon, krypton, and xenon.

Ramsay received the Nobel prize in 1904 after successfully demonstrating the radioactive decay of one element into another: radium into helium. Later he found one more element resulting from radium's radioactive decay: radon. Radon's atomic weight showed it to be the last in the series of rare gases.

41. Sir Henry Rawlinson—first to decipher cuneiform writing (1810-1895; English): The work of this archaeologist opened the way to a modern understanding of ancient Mideastern history. Rawlinson transcribed and successfully deciphered the great cuneiform inscription at Bisitun, written by Darius I. This feat provided the key to interpreting all the Babylonian-Akkadian inscriptions, just as the Rosetta Stone furnished the key to Egyptian hieroglyphics. Describing how Darius took control of the Persian empire after the insurrections that followed Cyrus's death, the Bisitun inscription corroborates the Bible's references to Darius. It also adds to our understanding of why the rebuilding of the Jerusalem temple was delayed at the beginning of Darius's reign until after the Persian empire had become stable. Rawlinson's excavations resulted in the discovery of many valuable Babylonian antiquities, which he donated to the British museum. He was also known as a committed Christian and Bible scholar.

42. John Ray—first to classify according to species (1627-1705; English): Georges Cuvier called Ray's surveys of birds, fishes, and insects "the basis of all modern zoology." Ray also published the first complete guidelines for classifying plants. By dividing flowering plants into those with single and those with double seed leaves, he established the basis by which these plants are still classified today. Modern botanists even acknowledge that his "natural" system, which made use of *all* of a plant's characteristics instead of just one, was superior to the "artificial" system of classification later used.

In matters of faith, Ray used the Bible as his authority rather than the Church of England. Thus he refused to sign the Act of Uniformity in 1662 and resigned his teaching post at Cambridge. From this decision, however, came the opportunity for his biological tours of Britain and Europe—and the basis of most of his knowledge for his later work. Ray wrote several books on natural theology, including *The Wisdom of God Manifested in the Works of the Creation.*

43. Bernhard Riemann—formulator of non-Euclidean geometries (1826-1866; German): This son of a Lutheran pastor pioneered several new branches of mathematics. Some idea of the extent of his influence on mathematics can be had by considering a few of the concepts and theorems that now bear his name: Riemannian geometry, Riemann curvature, the Riemann mapping theorem, the Riemann integral, the Riemann approach to function theory, etc. He is best known today for the fact that Einstein based his general theory of relativity on Riemannian geometry.

According to the geometry we are most familiar with (that of Euclid, Greek mathematician of the third century B.C.), the shortest distance between two points is a straight line. But Riemann proposed a geometry in which space is curved, and in which the shortest distance between two points is a curved line, or geodesic. In Euclidean geometry, the properties of space are defined the same everywhere; in Riemannian geometry, these properties might vary with local conditions. Einstein later borrowed this concept in his theory of gravitation to show how space-time is warped by mass. Reimann's mathematical concepts were influenced by the physical problems he dealt with while at the Göttingen physics laboratory.

Like his father, Riemann began his education with an intention to enter the ministry. An obvious prodigy in mathematics, he soon changed his course of study in that direction; but he always retained his love for the Bible, even writing of the mathematical evidences he saw of its inspiration.

44. Peter Mark Roget—forerunner of motion picture invention (1779-1869; English): During the 19th century, many people devised methods for viewing a series of pictures to create the appearance of motion. All owe their ideas to the fundamental concept first outlined in detail by Roget. In 1824, Roget presented a paper to the Royal society titled, "The Persistence of Vision with Regard to Moving Objects." Roget's early investigations led to the experiments of John Herschel and Michael Faraday and then to various commercial devices (which at first allowed "moving" pictures to be viewed on the rim of a rotating disk). Using drawings at first, the art of motion pictures thus predates the process of photography.

Roget had the knack of creating things that lasted long after he was gone. Trained as a physician, Roget helped to found the Medical School at Manchester and the University of London. His wide interests and varied accomplishments also took him into the field of mathematics and lexicography. Roget invented the "log-log" slide rule, universally used ever since. And, of course, he is especially remembered for his famous thesaurus, which groups words, not alphabetically, but according to ideas. The popularity and usefulness of *Roget's Thesaurus* continues through its new editions almost two hundred years after he began that work.

The son of a French Protestant pastor from Geneva, Roget's Christian convictions are perhaps most in evidence in his contribution to the Bridgewater Treatises, *Animal and Vegetable Physiology Considered with Reference to Natural Theology*. The fruit of those convictions is best demonstrated by the charity clinic he established (The Northern Dispensary), which he faithfully served as physician, gratis, for eighteen years.

45. Archibald Henry Sayce—world's foremost Assyriologist (1843-1933; British): Sayce's studies in ancient languages and texts dramatically increased our knowledge of the Assyrians, Babylonians, and Hittites. By the time he had become professor of Assyriology at Oxford, his reputation was already established as the final authority in questions of Assyrian grammar and comparative language study in the ancient Middle East.

Although Sayce had been ordained for the ministry at the age of 25, his studies had been infused with radical criticism, resulting in his low regard for the historical value of the Old Testament. During his own later archaeological and linguistic investigations, however, he learned that the theories of the higher critics had been developed without benefit of archaeological knowledge. His decipherment of text after text showed the close agreement of ancient Assyrian and Babylonian texts with the Bible's claims (see Volume 4).

His books, *Early History of the Hebrews* and *Israel and the Surrounding Nations*, raised many eyebrows among his peers. His celebrated scholarship, once used to challenge the Old Testament's history, was now used to vindicate it. Recorded contacts between Israelite and Assyrian kings were real; unheard-of nations who were once thought to be invented by Bible writers (like the Hittites) were shown to have been mighty empires. Sayce continued to write classic texts on Assyriology and contribute articles to *The Encyclopaedia Britannica* into the 20th century. Though Sayce was already an ordained minister, one of his chroniclers reports that Sayce converted to biblical Christianity later in life.

46. Sir James Simpson—founder of anesthesiology (1811-1870; Scottish): When the anesthetic properties of ether first became known, Simpson experimented with it in his obstetric practice. He read a paper to the Royal Medical society the following year, encouraging its use in certain cases. Simpson met with heated opposition from several quarters. Many warned of the dangers of this practice in obstetrics, citing medical and even Scriptural reasons. Having been trained in the Scriptures better than his critics, Simpson amply answered their biblical objections; and he continued searching for safer and more effective means of relieving his patients' pain.

After diligent and systematic experimentation, Simpson discovered the analgesic effects of chloroform in 1847. Once again he published a paper to describe and encourage the use of a new anesthetic, both in obstetrics and in surgery; and once again he was deluged with criticism. Surgeons, however, were soon impressed with the superiority of chloroform as an anesthetic; and when Queen Victoria insisted upon chloroform at the birth of Prince Leopold, all further criticism ceased.

Simpson made a number of other discoveries that contributed to the practices of gynecology and obstetrics. But his greatest discovery, by his own account, was "that I have a Saviour!" His search for anesthetics, according to some accounts, was inspired by the passage in Genesis 2:21, when "God caused the man to fall into a deep sleep."

47. Sir George Stokes—contributor to light and sound wave theory (1819-1903; British): This physicist/mathematician held the chair as Lucasian professor of mathematics at Cambridge while at the same time serving as secretary and president of the Royal society; he is the only person other than Isaac Newton ever to hold the three offices simultaneously. Through his investigations and his many papers (later collected in five volumes), Stokes advanced our knowledge of light, sound, hydrodynamics, and mathematics. Many of his experiments involved the transformation of waves while passing through various media. His investigations in hydrodynamics provided science with its best description of the suspension of clouds in the air and the flow of water in rivers and channels. Working with light, he furthered the wave theory of light and provided the first explanation of the phenomenon of fluorescence.

Stokes' Christian beliefs are amply displayed in his book, *Natural Theology*. These beliefs were also borne out by his reputation for selflessness. While his friends claimed that credit for Kirchhoff's theory of spectrum analysis should go to Stokes, since his discoveries were earlier, Stokes himself maintained that Kirchhoff's theory was fuller and that Kirchhoff deserved priority. Stokes also took his role as professor seriously; all students knew they were welcome to come to him for help with their studies—and colleagues continued to seek him out for counsel. By his unselfishness, Stokes promoted good science—not only through his own research, but by the ingenious problems he posed for others to investigate.

48. Thomas Sydenham—discoverer of malaria's cure (1624-1689; English): Sydenham's discovery of quinine was no fluke. Others had tried the drug (from the bark of the cinchona tree) and given up on it when its use seemed to help some patients and speed the deaths of others. Sydenham rightly reasoned that the initial relief from the drug could be prolonged if it was administered in proper amounts, in the proper cycles of the disease, and at proper intervals after initial symptoms disappeared. His cautious experiments led to the right answers in each of these matters. He also had faith in a God who had designed nature in such a way that the physician's task was to help nature complete the cure. "A disease," he wrote, "is the effort of nature to restore the health of the patient by elimination of morbific matter." With this conviction, he correctly decided to wait until after the initial malaria attack before administering the quinine.

In a day when mystical arts had crept back into medicine,* Sydenham, more than anyone else in the 17th century, helped to bring scientific methods into medicine. His close friendship with Robert Boyle may have played a part in his experimental approach to his work. Sydenham and Boyle had in common their faith in the scientific method as well as their faith in Christ. The son of a Puritan, Sydenham was not disposed to glibly accept traditions. "It is my nature," he wrote, "to think where others read; to ask less whether the world agrees with me than whether I agree with the truth." Such independent thinking also led him to make great advances in the treatment of measles, smallpox, gout and many other diseases. No one before him had ever set about to analyze and describe the conditions of each disease. His careful observations laid the groundwork for the later scientific understanding of infectious diseases and their successful treatment.

49. Theodoric of Freibourg—discoverer of the rainbow's cause (c. 1250-1310; German): Theodoric not only gave the world its first accurate explanation of the rainbow's formation; he also gave the world its finest example of experimental science during the Middle Ages. In an age when the scholastics narrowly adhered to their traditional authorities and when science consisted of repeating the works of Aristotle, Theodoric struck out into new investigations, using empirical methods that were not to become standard until centuries later.

Theodoric observed the effects of light passing through water-filled glass globes in order to empirically test his theories about how raindrops form the colors and shape of the rainbow. Through his many experiments, he succeeded in showing that colors were formed in the interaction with the water drop, not merely in the eye of the beholder. Theodoric's careful investigations with glass globes even showed that the particular order of colors could only be the result of light being refracted at the raindrop's surface and then reflected deeper inside the raindrop. He could produce a second rainbow, with the order of colors reversed once more, by creating a second internal reflection. Thus he also explained how secondary rainbows often form, with the order of their colors reversed. Theodoric showed how the phenomenon of colors can be explained by differential refraction. That is, each color in white light is refracted differently at the raindrop's surface, causing the differently colored rays of light to travel different pathways within the drop.

Theodoric chose to follow his Lord as a member of "the Order of Preachers," the Dominicans. Even here he broke new ground, for he was

---

* To heal a wound, some Elizabethan doctors (just one generation earlier, during the days of Paré, q.v.) had made it a practice to apply ointment, not to the injury, but to the weapon that produced it.

the first educated man ever to preach to his people in German, their own language. Once again we see a scientist who, while living out a personal faith rather than merely going through the motions of religion, brought the same independent spirit into the scientific realm.

50. William Whewell—inventor of the anemometer (1794-1866; British): Whewell's invention of the modern version of this device for measuring wind velocity is merely one of his numerous scientific contributions. As a philosopher of science, he is best known for his promotion of scientific inductive methods and his invention of dozens of scientific terms now in common use. Whewell first formulated the terms *scientist, physicist, anode, cathode, electrode, ion, electrolyte,* and the geological epochs: *Miocene, Pliocene,* etc. An ordained clergyman, he wrote and taught of the wisdom of God seen in nature. As his contribution to The Bridgewater Treatises, Whewell wrote *Astronomy and General Physics Considered with Reference to Natural Theology.*

A large, highly inclined spiral galaxy, NGC 253. —*Courtesy of Lick Observatory, University of California*

Everything You Ever Wanted to Know About 20th-Century Cosmology

Timeline of 20th-Century Cosmological Discoveries

1905   Albert Einstein publishes his theory of special relativity, introducing the concept of space-time and describing bizarre phenomena when the speed of light is approached.

1912   Henrietta Leavitt, of the Harvard College observatory, discovers how the period of Cepheid variable stars is related to their absolute brightness (a discovery that later provided others with a tool to measure distances to galaxies containing Cepheid stars).

1914   American astronomer Vesto Slipher reports that almost all nebulae are redshifted, that is, receding from us at high velocities.

1915   Albert Einstein publishes his equations of general relativity, describing gravity as the effect of mass on space-time. Newton's 300-year-old theory of gravity is finally replaced.

1917   Dutch astronomer Willem de Sitter finds a solution to Einstein's field equations that implies that the universe is expanding.

1919   English astronomer Sir Arthur Eddington leads an expedition to the Gulf of Guinea (west Africa) to photograph the total solar eclipse. By finding that starlight is bent as it travels past the sun, he provides the first verification of Einstein's theory of general relativity.

1922   Russian mathematician Alexander Friedmann shows that Einstein's field equations do not allow a static universe.

1923   American astronomer Edwin Hubble finds a Cepheid variable star in the Andromeda nebula, indicating its tremendous distance and that the nebulae might be distant galaxies in their own right, not objects within our own Milky Way.

1927   Belgian physicist Georges Lemaître proposes his theory (now known as the big bang theory) that the universe is expanding from a "primeval atom."

1929   Hubble announces his finding that all distant nebulae are retreating from us with a velocity that is related in a linear manner to their distances.

1929   English quantum physicist Paul Dirac proposes the existence of antimatter, showing that matter and antimatter must result in equal quantities from the transformation of energy into matter.

1932   Bell Telephone Lab's Karl Jansky discovers natural radio signals coming from the center of the Milky Way galaxy. He invents the radio telescope, making all later deep space findings possible.

1932   Carl Anderson, of the California Institute of Technology, first detects an antimatter particle, the positron, in his cloud chamber. He photographs the vapor trail of a particle that takes the path of a positively charged particle, though it has the same mass as an electron.

1948   George Gamow, Ralph Alpher, and Robert Herman use data from atomic experiments to show conditions in the intense heat of the early universe, before atoms could form. Their theory shows that, as a result of the big bang, about three-fourths of the universe should be composed of hydrogen and one-fourth helium. They also predict the existence of an afterglow from the fireball: leftover background radiation that should have now cooled to about 5 degrees Kelvin everywhere in space.

1950s   Fred Hoyle, William Fowler, and Geoffrey and Margaret Burbidge show how heavy elements (composing less than 1 percent of visible matter in the universe) result from cooking hydrogen and helium in stars.

1950s   Oxford radio astronomer Martin Ryle finds that radio galaxies become more abundant with distance, meaning that galaxies are distributed more closely the further we look back in time.

1955   Emilio Segré and Owen Chamberlain use a particle accelerator at Lawrence Berkeley Lab to generate antimatter. Protons are shot at other protons, producing pairs of protons and antiprotons.

Early 1960s   Allan Sandage discovers quasars; others show that the extremely high redshifts of these powerful objects tell us of an early universe that was very different from today.

1963-1973   Brandon Carter, Stephen Hawking, Werner Israel, Roy Kerr, Roger Penrose, David Robinson, and John Wheeler all make contributions to the theory of black holes, concluding that anything that collapses to form a black hole will end up looking like the same thing; its size and shape will vary only with its mass and rate of rotation. The presence of black holes thus becomes more predictable by our increased ability to make detailed models to compare with observations.

1965   Bell Lab's Arno Penzias and Robert Wilson detect the microwave background radiation at about 3 degrees Kelvin everywhere they point their specially cooled, sensitive horn antenna.

1967   Russian physicist Andrei Sakharov proposes that the big bang somehow produced a just-right excess of matter over antimatter, resulting in the annihilation of antimatter and forming the universe of matter we see today. This would explain why antimatter has not been found in any quantity in our universe.

1970s   Robert Dicke and James Peebles announce their finding that, as they put it, Omega must be near 1. That is, there is a critical density for matter in the universe, and if our universe's actual density is more or less than this amount, the universe should have either collapsed a long time ago or should have been too dispersed to allow galaxies to form.

1974   Cambridge University cosmologist Brandon Carter proposes "the anthropic principle" to explain the precision of physical laws and conditions that make life possible.

1976   George Smoot, of Lawrence Berkeley Lab, uses his differential microwave radiometer in a high altitude U-2 spy plane to detect dipole anisotropy in the microwave background radiation. This dipole, or temperature difference between opposite directions, shows that our galaxy is being dragged at more than a million miles an hour by a great unseen mass, and that clusters of galaxies must form structures larger than previously imagined. The beginnings of such structures should have left small ripples in the microwave background radiation.

1979   Alan Guth publishes the first inflation model of the big bang theory.

1980s   David Schramm of the University of Chicago proposes that at least 90 percent of the matter in the universe is "dark matter."

1986   Valerie de Lapparent, Margaret Geller, and John Huchra publish their findings that galaxies not only form superclusters in "bubbles," but that some of these structures are hundreds of millions of light-years in extent.

1992   George Smoot announces the COBE satellite's measurement of the predicted ripples in the cosmic background radiation.

1993   John Mather reports that his FIRAS instrument (far infrared absolute spectrophotometer) on the COBE satellite measures the deviations between the cosmic background radiation and a perfect blackbody radiation at less than .03 percent.

Q & A with
Today's Leading Cosmologists

When a skeptic reads about 20th-century cosmology for the first time, he naturally asks: How do they *know* that? How do they know the universe is very near the critical density required for life? How do they know the universe is expanding? Why are they so sure this microwave background radiation is the remnant of a creation event? In order to focus on my main points, I did not always allow my interviewees to tell their whole story. And I didn't deal at all with what these astrophysicists typically think about the relationship between science and faith. But for the sake of the thoughtful skeptic who wants to dig a little further, Searchlight, Inc. offers the following crash course in astrophysics and the philosophy of science—taught by the scientists who have made some of the greatest contributions to modern cosmology.

Where Science Leaves Off

Astronomer Robert Jastrow (founder of NASA's Goddard Institute and now serving as head of Mount Wilson's observatory) had this to say when I pressed him to tell me about anything science could tell us about how something could come from nothing.

Heeren: Is there anything we know now from quantum mechanics or inflation theory or anything else that can explain how the universe—and space itself—could have come out of absolutely nothing?

Jastrow: No, there's no—this is the most interesting result in all of science. Whether they came out of nothing or out of a pre-existing universe, as a product of forces that we have never discovered, no one knows the answer to that question, because the circumstances of nearly infinite heat pressure and density at the beginning of the universe necessarily wiped out any trace of a previous universe. So time, really, going backward, comes to a halt at that point. Beyond that, that curtain can never be lifted. . . .

As Einstein said, scientists live by their faith in causation, and the chain of cause and effect. Every effect has a cause that can be discovered by rational arguments. And this has been a very successful program, if you will, for unraveling the history of the universe. But it just fails at the beginning. And that is really a blow at the very fundamental premise that motivates all scientists.

As both a working scientist and a Christian apologist, Robert Gange brings a unique perspective to the issue of where science ends and faith begins. Dr. Gange serves as president of the Genesis Foundation in New Jersey, an organization chartered to show that the Bible is trustworthy from a scientific perspective. As the author of the book, *Origins and Destinies*, he argues that Christians must be careful not to try to cross-fertilize the separate realms of science and faith. And science, he told me, has no business trying to deal with questions about ultimate origins.

Gange: Science is, in my view, a Ph.D., meaning P for predictions, D for data, and H for hypothesis. Those are the three essential elements of science. Now the predictions have to be logical, the data have to be reproducible, and the hypotheses have to be falsifiable. . . . So what happens here in science is that observations are made, and data is acquired through those observations that are replicated through other workers elsewhere. And so in that sense, the data is accepted as real because other people see it also. So it's a collective enterprise.

Then, that data inspires a hypothesis. There's a hypothesis that comes up to explain what one sees. That hypothesis in turn begins to produce predictions as to what else ought to be if the hypothesis is true. So then additional data gathering inspired by the hypothesis occurs. And again, it's reproducible. And so long as the hypothesis continues to answer every question that we can think to ask of it, it remains true. The moment it cannot be altered to satisfy what we're asking of it, it's falsified.

I've really quickly gone over this, but the fundamental point that I'm trying to bring out is, nowhere in that enterprise does the word *faith* appear. This is a collective judgment on the part of an international community of people who are trained in various disciplines and who meet on a regular basis in different ways to basically create some sense out of what is seen and understood. Faith is an entirely different issue. . . . In my earlier years, I used to say, "Yeah, you can cross-fertilize those two." But as I've grown older, I tell you, in the last analysis, faith is truly a thing separate and apart from science. Now I will say that science is a friend of Scripture and not a foe. And especially when you get to the question of origins and the question of destiny, science has no place there. . . .

There are only three categories of events in space and time: events that are reproducible, events that are unpredictable, and events that are singular. An event that's reproducible is the data of science. If you, for example, want to measure how fast an apple hits the ground when released a certain height above the ground, you can repeat that over and over and over again. . . . So reproducible events lend themselves to scientific inquiries. Whereas unpredictable events lend themselves to statistical inquiries. Events that are singular lend themselves to legal inquiries. And the question of whether there was a creation of the world or there is a destiny to man are singular inquiries. These are singular events. The creation of the world is a one-time event—it's a legal inquiry. Science has no proper jurisdiction in the question of origin or destiny.

To me, science is useful in helping us see that biblical faith can be reasonable, that it is not reserved for people who are ignorant of the real world and how it works. I have not written this book to *prove* the Bible. Nobel-prize winning physicist Arno Penzias, however, appears to take a further step in separating science from faith:

Penzias: I do not believe that anyone should ever say that science agrees with religion. What I would say, which I think is a far more powerful statement, and one which allows people to be religious, is to say, the modern observations of science do not *disagree* with

Robert Wilson (left) and Arno Penzias shortly after receiving the Nobel prize for physics in 1978. In the background is the antenna that first detected the microwave background radiation in 1965. —*Courtesy of AT&T Bell Laboratories*

religion. . . . The double negative, the "not to disagree" I think is far more powerful, because after all, if tomorrow we find the steady state theory is right, would that mean you'd stop giving to charity?

Heeren: Good point.

Penzias: We're merely saying that if science for a while goes into disagreement, we don't lose our faith. If we go back to Maimonides' *Guide for the Perplexed*—it's a traditional, Jewish text, which says . . . I think it's, "Do not trust the words of any man [meaning particularly Aristotle, who taught an eternal universe], for it is the foundation of our faith that God created the universe from nothing, and that time did not exist before."

So what Maimonides said was that just because science is inconsistent at the present time with the tenets of our faith, don't believe it. So in other words, we are saying we can't have it both ways. We can't have scientists say that when there's disagreement, as there was in the days of Aristotle, that therefore we should stop believing. . . . Our ancestors had enough faith to believe even when science happened to be against them. I think it demeans religion to attempt to strengthen religion with this agreement. . . . I think instead what we ought to say is, "I think it's significant that there's no disagreement." I would hate to have a religion which is based on science. . . . It's a double negative. In my case, what I will always say is what we find is there's an absence of inconsistency.

Heeren: Couldn't I turn that around just as well though and say the findings of the 20th century *are consistent* with the concept—"

Penzias: No. That's very different. That's *very* different.

Heeren: Really.

Penzias: I think the absence of inconsistency is very different—to me—from consistency. I think there's a subtle difference, and I would insist on the difference. Because consistency between the two implies a much larger congruence. . . . The scientific perspective is a limited one. And because it's so limited, I think it's really that in this little part we find the absence of disagreement is much better than the agreement.

Curious to know if this was a commonly held view for scientists, I asked Dr. Jastrow about this distinction.

Heeren: Arno Penzias told me that he would never say that any finding from science is consistent with the concept of a Creator, but rather he would only say that a finding is *not inconsistent* with such a concept.

Jastrow: Yeah, that's a very funny turn of language of my physics colleagues. "Not inconsistent with" is an identical synonym for "consistent with."

Heeren: Well, that's the way it struck me, but he wanted to make that very careful distinction.

Jastrow: They all do, but there isn't a distinction. But I think what he wanted to say was that the presence of an intelligent being, a Creator, a Designer, was independent of our physical knowledge, that science could not illuminate the question of such an entity existing.

Heeren: Do you feel that this extreme carefulness about avoiding all talk of a Creator is a necessary one for good science, or that some scientists use it as an excuse to avoid thinking about God?

Jastrow: They tend not to be philosophically inclined. Maybe the point is that they feel instinctively that such a question is not answerable within the limits of their field. So since their basic understanding of the world is that everything should be answerable, they just don't like to talk about it.

When I talked about the consistent/not-inconsistent distinction with George Smoot, discoverer of the famous ripples in the microwave background radiation, he told me:

> When scientists say something is consistent, they say, "The theory says that the earth is round, and I went out and measured an eclipse and I saw the shadow of the earth go across the moon, and it was round to within one percent, it's consistent with the theory that the earth is round." See, consistent, for scientists, means it's very supportive.

This helped me to see that, in the language of scientists, to say that the God of the Bible is consistent with science is practically the same as saying that science *proves* the existence of the Bible's God, as if science has found evidence to prove 99 percent of all that the Bible tells us about Him. Obviously, if science were that powerful, it would leave little room for faith. Faith would undoubtedly be easier if science had such power, but if faith became as common as dirt, could it then be as valuable a thing in God's eyes? This brings us back to the point Margaret raised in the opening story of this book. The faith that pleased Jesus most was evidently a rather rare thing (Luke 7:9; 18:8). There may be wisdom in listening to the admonitions of Arno Penzias and Robert Gange, who both see the need to clearly separate faith and science.

So of what use, from a spiritual viewpoint, is all this exploration into the findings of science? First, if the data of science do not contradict the Bible, then a skeptic can no longer use science as an excuse to reject the gospel. This book/tape series has not been produced to encourage anyone to build his faith on science. At the same time, the findings reported in this series could help remove some major stumbling blocks to faith. After all, many have been taught (or have assumed) that the findings of science are *inconsistent* with the Bible. This needs to be cleared up. For the Bible believer who wishes to reach the skeptic today on his or her own terms, this may be an essential step before the skeptic will even begin to take the Bible's message seriously.

Further, what we *can* say is that the discoveries of modern science *are consistent* with a creation event and with intelligent design. Atheism and most of the world's non-biblical religions are *inconsistent* with modern cosmology; and during the periods of the Bible's composition, *all* other religions were inconsistent with modern science. Those who first believed in the God of the Hebrew Scriptures and the Christian New Testament stood alone against the views that surrounded them. They had neither the ability nor the need to have their beliefs "vindicated" by 20th-century cosmology.

Today, I maintain that the greatest scientific discoveries of our century make great conversation starters for the Christian witness; it makes good sense to begin with facts that both believers and unbelievers can agree upon. And then, the separate realms of science and faith can actually be used to advantage. To begin to share one's faith, one needs only to show how modern science has raised questions it can't answer, or as Robert Jastrow might put it, that 20th-century science has led us to a curtain that science can never raise. This is where science ends. And this is where the Bible begins.

How Do We Know Our Universe Is Very Near the Critical Density?

Penzias: If I put on my physics hat, I thinks it's so unlikely as to make me shudder as a scientist. On the other hand, as an astronomer, boy, all I have to do is look at the parameters. If this had been anything but galaxies in the universe, everyone would have taken the data at face value long ago and said the universe doesn't have enough matter to pull it together. There'd be no reason.

Evidently the existence of galaxies takes some explaining, and science has had to scramble for theories that have no basis in observations in order to explain how our universe might possibly have reached very close to the precise density required for our existence. Though scientists continue to devote themselves to involved theories involving inflation and exotic matter, neither theory yet has a shred of observational evidence to suggest its reality.

Calculations easily tell us that the likeliest scenario should either be a universe that collapses too soon or disperses too fast to permit life. Any other result has a likelihood of one in billions, by anyone's estimate. Yet actual measurements tell us that the universe has very close to the required mass for critical density (at least a tenth or so of the needed amount). And as we have seen, this means that the ratio between the universe's actual density and the critical density had to be either 1 or within billionths of 1 at the very beginning. But how do we come up with these actual measurements? How, the skeptic might ask, do astronomers even begin to "observe" the amount of mass in the universe? For this we should turn to one of Princeton's renowned observational sticklers, Jeremiah Ostriker.

Heeren: How does one measure the amount of mass in the universe?

Ostriker: The way I like to think about it is this: suppose you want to know what you weigh. You can say, "Well, my pants are getting tight, so I guess I put on weight. Or I look in the mirror or I ask my friends what they think. Any number of different ways of doing it. But there's really only one real way to find out. And that's standing on the scale. There's only one way to find out what things weigh and that's through gravity. It's the *only* way. Everything else is indirect. Okay?

And so you use gravitational measures. And an example that we're familiar with is we get the mass of the sun by the orbits of the planets around it. It's basically their velocity squared times the distance divided by Newton's constant of gravity. This gives you the mass of the sun. Well, we do the same thing for galaxies and groups of galaxies and clusters of galaxies, etc., etc., and we find the mass that way. And that, I've always believed, is the only way.

Sometimes people use the light, and they take the observed light times the mass to light ratio for the object, and they get the mass that way. But that's an observation times an assumption, if you wish, because who knows what the mass to light ratio is?. . . . So the answer is that you should always use gravity, and it always involves velocities and distances and Newton's laws.

When I asked Nobel laureate Robert Wilson about how we know that the density of the universe is very close to the critical density, he gave me a good summary of the modern scientific consensus. After explaining how the mass of galaxies is measured in similar terms that Ostriker discussed (using gravity), he went on to discuss the need for dark matter.

Wilson: When you do this [apply the laws of gravity], you find that there is more mass than you can account for—from the mass to light ratio. And that's the first evidence for dark matter. If you were to simply take the amount of light that is observed in galaxies and sort of add it up over the universe, you come out with about one-hundredth of the critical mass. . . .

Heeren: So this dark matter—it's not a question of whether it's there— it's a matter of what it *is?* No one disputes the fact that it's there?

Wilson: Right. I know one other measurement that has been made other than just the dynamics: that is that Tony Tyson looked at the gravitational deflection of light—from a quasar, I believe— in a very distant cluster, and came out again with the requirement for dark matter in that cluster. So in addition to the gravitational attraction of the galaxies in the cluster, there also was the effect on light passing through one. So if you put in all of the dark matter, you're missing sort of a factor of ten in closing the universe.

Heeren: Okay.

Wilson: Now when you go back to general relativity and look at the expansion, what you find is if there's a tiny deviation on either side, it will exponentially depart from critical density. So if it started out with one part in 10 to the 60th too little mass, there wouldn't be stars and galaxies and planets and things—we just wouldn't be here to observe it. And if there were the same amount more, it never would have expanded to this point; it would have re-collapsed. So the requirement after inflation is very exacting, as to the density of the universe, to come out near the critical density. . . . But the observations put us within a factor of ten now, and therefore within some *very* tiny fraction, back at the earliest time we can think about.

What Do Leading Cosmologists Think About the Anthropic Principle as an Explanation for Design?

The critical density, of course, is just one of many critical parameters that demand some explanation. Most popular texts (including this one) take quite a bit of space to talk about the only explanation naturalists have for these mysteries: the anthropic principle. But to get a fuller perspective on the subject, we should also examine what scientists actually think of FAP,

PAP, SAP, WAP, etc. How seriously do working cosmologists take the concepts of humans becoming God, of humans observing themselves into existence, of multiple universes, or even the idea that humans simply occupy a privileged time and place?

"Oh, yes," Robert Jastrow told me, "I talk about that anthropic principle, but I don't give any much credence to it."

George Smoot told me: "I do not have confidence in its use. It has little predictive power."

Jeremiah Ostriker said: "I think they're very interesting arguments. And myself, I've never known what to make of them. So I find them interesting but not compelling."

As I mentioned earlier, Alan Guth believes that the problem with the anthropic argument is that it can be used to explain anything. He told me: "My point of view is that the anthropic explanation is always the resort of last recourse. If you can't find any intelligent theory that's compatible with what you see, that will predict what you see, then you might, as a last resort, entertain a purely anthropic explanation."

And last, the most interesting opinion came from the esteemed Lucasian Professor of Mathematics at the University of Cambridge, Stephen Hawking. He agreed with the rest, but his reasoning was surprisingly different:

Heeren: Much of your work is apparently aimed at finding a better explanation than the anthropic principle for the universe we observe. Why do you find the anthropic principle inadequate as an explanation for this century's findings?

Hawking: The human race is so insignificant, I find it difficult to believe the whole universe is a necessary precondition for our existence. Clearly the solar system is necessary, and maybe our galaxy, but not a hundred billion other galaxies.

I was surprised that Hawking answered this way. The *extent* of the preconditions for our existence is not really the question. Naturalists must deal with the problem of explaining the many "unnatural" selections whether they are considering the entire universe or just the solar system. The odds against any of the nine "unnatural selections" I described in Chapter 9 apply as fully to the solar system as to the entire universe (the existence of our life-conducive elements, the ratio between the masses of the proton and electron, the relative strength of the four fundamental forces, etc.).

Even 19th-century scientists (without knowledge of the far-off galaxies) asked themselves, "Why is our world so perfectly suited to our existence, as if designed just for us?" When Hawking answers that our galaxy may or may not be "necessary" for our existence, but that the solar system clearly is, he still leaves open this real question. Why *should* our solar system, or any other part of our universe, so clearly meet all the many critical requirements for human life against incredible odds?

As Barrow and Silk pointed out, the universe has to be as enormous as it is to "support even one lonely outpost of life." But why would the entire universe meet so many critical preconditions against incredible odds just for us? Why would anything other than God go to all the trouble?

Hawking's reference to billions of other galaxies actually helps us to see the absurdity of any naturalistic system that would take such exacting pains to give birth to an entire universe of such dimensions, all for the sake of a seemingly insignificant race such as ours. In his writings, Hawking makes it clear that, like most scientists, he objects to both the weak and strong anthropic explanations on the grounds that they have no real value as explanations or basis in science.

Quantum Mechanics and the Biblical Picture

Those scientists who say that science may one day unlock the deepest mysteries of existence believe that a deeper understanding of quantum physics holds the key. Quantum mechanics raises a host of knotty questions. Do our observations put humans back in the center of the universe? Does observation actually create anything? What are the deepest implications of quantum mechanics on the kind of world we live in? Does nothing really exist but thought?

Having a keen interest in quantum physics, physicist Robert Gange readily gives answers to each of these questions: Yes, no, a world that must be continuously sustained by God, and sort of. Here's a fuller treatment of his thought-provoking explanations of the deepest implications of quantum mechanics.

Gange: The deepest conclusion of quantum physics is that the only reason anything has physical existence is because of human consciousness. In a sense, man has been reestablished as the center of all creation. . . . All scientific laws today are really statistical averages of quantum laws, which are, in point of fact, descriptions of human observations. And so it's no longer *things* that are described by science, but *observations* that are described by science. . . . So what in point of fact *is, is* only because of human consciousness that is making observations through this particular oxygen-burning organic machine. . . .

Heeren: How do you feel about the anthropic arguments as an explanation for what otherwise appears as design?

Gange: What you're really saying is what quantum physics has already concluded: that human consciousness is primary.

Heeren: But then can't an unbeliever say that man has actually created himself by observing himself?

Gange: No, not at all. That's not an argument that says that man has created himself. It's an argument that says that in creation all of creation *is* because man exists. It has nothing to do with man creating. . . .

Actually, to get biblical with that: The biblical picture of creation is not Greek substance philosophy. In Greek substance philosophy, the world gets put into existence and it can stay in existence of its own accord. That is not the biblical picture. The biblical picture is that all things hold together in Christ—Colossians 1:15-17 and Hebrew 1:3*—that all things are upheld by the Word of God. So the biblical picture is that creation is an ongoing theater of events that owe their coherence and existence to the continuous uplifting power of God—continuously.

Imagine a big color television set. You have signals coming into this antenna. It's not so much, were those signals to stop, that the picture would be confused, as it is that there would be no picture at all. This whole physical creation is that television picture, and the biblical picture is that there are constant signals coming in. Okay?

Heeren: Okay.

Gange: All that quantum physics is teaching is that without human consciousness nothing exists *as far as we know from our science.* But the Bible's saying that it does exist—it exists from God, and the only reason now quantum physicists are saying it exists is because we have life, consciousness. So that takes you away from the eastern, mystical malarkey that people try and get us into. It's got nothing whatever to do with creating reality. It has to do with the fact that the biblical picture is that creation exists because of the uplifting, continuous power of God, and that human consciousness is the mirror that allows us to affirm its existence.

---

*Hebrews 1:3 says that "the Son . . . is sustaining all things by His powerful word."

Did the Big Bang Really Happen?

Though James Truran's specialty is early galaxy formation, he makes no effort to make the newest Hubble Space Telescope galaxy photos conform to the big bang theory. Truran (University of Chicago) is known for being very reserved about his interpretations of the farthest galaxies we can now observe. Among astrophysicists, he seemed like a good choice to give an objective assessment of the real evidence for the big bang.

Truran: An enormous success of the big bang theory is the microwave background radiation, which is not easily explained in other ways. Maybe it's explainable—some people would argue, yes, of course it's explainable, but they very often have to make a set of assumptions rather than a single assumption to make this explanation hold together.

Heeren: Particularly that blackbody curve, I assume, is difficult to explain any other way.

Truran: The beautiful thing from COBE was the magnificent uniformity of this blackbody curve—just spectacular. I've had people say to me, "The enormous homogeneity of that blackbody is in some sense more spectacular than the fact that COBE also found evidence for anisotropy [that is, the fluctuations were uneven] at this very low level.

Heeren: And so those fluctuations may not be as significant as how close it comes to the true blackbody curve.

Truran: No, they certainly are, because we were looking and saying to ourselves that there must certainly be this level of inhomogeneity at some point—we wanted to find where it was, and certainly we were excited for it to be there, but also the blackbody curve itself—the beautiful curve they got—was at least an equally spectacular result of COBE.

And then superimposed upon that—the microwave blackbody radiation—is the fact that the standard model seems so well to predict the observed abundances of the isotopes of hydrogen, deuterium, helium 3, helium 4 and lithium 7. It seems to hold together. Observations of the oldest stars in our galaxies indicate a level of lithium 7 which is really right there where it has to be in the framework of this theory. And so it's this kind of strength to the theory that we're looking for. We want a theory which doesn't involve too many assumptions that seem quite exotic, which at the same time can explain an increasingly large number of observational, known features of the universe as we observe it. That's its strength.

Does everyone agree that the microwave background has its source in the big bang's fireball? And does everyone agree with the importance of finding the ripples in it? According to Arno Penzias, the microwave background stands on its own. We never needed to find the ripples to prove its source.

Penzias: Now as far as George Smoot is concerned in this matching ripples with the big bang theory, I think it's rather overblown. After all, nobody had ever explained away the microwave background radiation, and at some level one had to find ripples.

Heeren: So you're saying it was very well established before that and—

Penzias: Sure, the microwave background radiation was well established for a long time. And it's very nice that people found the ripples. But certainly, I don't think anybody but doubted that the microwave background radiation was what it was, and [there was] no other explanation.

Robert Wilson described the process of looking for other explanations at first, and the consensus on other alternatives today:

Wilson: Over the years, there were several attempts to reconcile this— the existence of the background in the steady state universe—all of which failed, I think. If you asked Fred Hoyle, he would probably say right now that, no, there's an easy explanation, because he still believes—I think he has come back to believing the steady state and that he knows how to generate the microwave background.

Heeren: I think it's through a quasi-steady state that actually involves multiple big bangs or medium bangs, isn't it?

Wilson: Yes, that was at least one of his efforts. I think that no one believes that his mechanism can match the precise thermal spectrum that COBE observed. He may still, but I don't think anyone else who has seriously looked at it does.

Stepping up to the chalkboard, Alan Guth explains just why astrophysicists would expect to find cosmic background radiation coming to us from the time when the universe first became transparent, about 300,000 years after the big bang:

Guth: Before a few hundred thousand years after the big bang, the universe was so hot that it was filled with a plasma—there weren't neutral atoms, the atoms were ripped apart—so the electrons moved around freely from the nuclei. And when matter is in that plasma state, it's incredibly non-transparent to radiation. . . . The free charged particles interact very strongly with the photons that make up light. So the photons are constantly being scattered and absorbed—they never really get anywhere.

Then, according to our calculations, the universe cooled at about 300,000 years after the big bang enough so that the matter rather quickly became neutral, neutral atoms. So it was a gas, like the air that we have around us. And a gas of that sort is very transparent to light, which is why we can see each other. And it means that the typical photon of this cosmic radiation has in fact been traveling essentially on a straight line since about 300,000 years after the big bang.

So when we look at this cosmic background radiation, we're really seeing an image of what the universe looked like at about 300,000 years after the big bang. Now, if it were completely smooth then, we would not be able to understand how galaxies ever formed. In order for galaxies to form, you need to have slight excess mass in some places to produce a gravitational field which then pulls the matter in. It's an unstable process. Once mass starts to collect in a certain region it produces a strong gravitational field and pulls in more mass, but in order for that to get started, you do need some nonuniformities that must have existed in the early universe. . . . And it finally was seen by the COBE satellite, the Cosmic Background Explorer, in, I guess, 1992, and since then there have been confirming observations made from ground-based and balloon-based experiments. And it's very exciting, because we're still exploring the details of that angular pattern.

How closely do the ripples in the microwave background actually match the expected ripples of the big bang theory? Direct from NASA's Goddard Space Flight Center, John Mather gives us his direct answer:

Mather: Well they match just exactly as well as they should. In other words, given the uncertainties in the measurements, they fit perfectly.

The leader of the team that discovered the ripples is more coy:

Smoot: That depends on who's theory it is. The big bang has several different versions.

Heeren: That was going to be my next question: Which model seems to have the most evidence right now?

Smoot: Well, the model that most people go for is the inflationary model, and even in that there are many versions of inflation. But most of them predict that there'll be a power spectrum of fluctuations that will be scale invariant—that is, the number or the fraction you get is independent of scale. And so in terms of that, are the data coming up in the right level or not? And the answer is they *are* in fairly good agreement with that. The question though is, "Can we distinguish between the various models that are viable?" And the answer is "No, the measurements are not good enough yet."

Heeren: But the measurements are good enough to help to—they are consistent with a big bang theory.

Smoot: Right, they're consistent with the big bang theory, and with the idea that galaxies and so forth formed from gravitational instability. But it's not clear that we can distinguish between the various models right now. But on the other hand, there are other models that are ruled out, although you keep hearing about them.

Has Inflation Eliminated the Need for Fine Tuning?

To find out, I asked Alan Guth whether the newest inflation models have successfully avoided having to assume that any fine tuning went on in the matters that concern the inflationary big bang.

Guth: As far as finely tuning things, there are still two important fine tuning problems that are not solved. One is the problem that's called the cosmological constant problem. It's basically the problem of why the energy density of the vacuum is either zero or very close to being zero. Current models of physics require fine tuning in order to make the energy of the vacuum turn out to be either zero or very, very small.

Heeren: I see.

Guth: And that's not understood. That's a basic problem with particle physics. The second problem is more directly related to inflation. The cosmic background radiation is uniform in temperature to about one part in a hundred thousand. In order to get these nonuniformities to be as small as what we observe, we have to arrange that certain numbers that describe the underlying particle physics be very, very small, for reasons which we do not, at the present time, understand. So that's another instance of fine tuning, which is not yet overcome by any of the theories that we have.

Evidence for a Supernatural Creator

Looking at the discoveries of modern science, Robert Gange finds powerful evidence of a Supernatural Creator. But he doesn't start his argument with the discovery that the universe must have had a beginning or with the evidence for design. He starts with the evidence that the universe is *old*.

Gange: The thing that argues for the existence of a Supernatural Creator is the fact that the universe has been in existence for between 14 and 17 billion years. Now that almost sounds contradictory. Most Christians who are trying to argue the Henry Morris line are trying to say that everything is very, very young. What they're not realizing is the fact that scientists today accept ages of the order of 14 to 17 billion years is itself proof of a supernatural creation.

Heeren: How so?

Gange: Because all of these models today have one common denominator, and that is they all teach that the universe should have gone out of existence in an infinitesimal fraction of time after it came into existence. That time is called a Planck time, and it's 10 to the minus 43 seconds.

Heeren: Right.

Gange: Now, since the universe should have gone out of existence in under a Planck time, and since it's 14 to 17 billion years old, according to scientific thought, the conclusion is it must have been tuned at inception to better than 60 decimal places. All natural processes have a maximum precision; 3 or 4 decimal places is enormous. To say 60 decimal places is to literally genuflect at some supernatural creation.

These Christians who put a straight jacket on God, who try and insist that God couldn't have done something a certain way, to me is just incomprehensible.

Heeren: To me there are two big bodies of evidence that we can point to: one is the whole idea of a big bang that you just described, and the fact that we're still here after all this time. The second would be the many numerical coincidences that people like Hawking say are "very finely adjusted" or "carefully chosen," like the ratio between the masses of the proton and the electron—

Gange: You have to be careful there because, again, all you're nailing it to is a probability argument. You know what the counter argument is?

Heeren: Multiple universes?

Gange: Yes, the counter argument is, "Well, we could have had billions and trillions and skillions of universes and this just happened to be the one that had the right numbers." I mean, there's no end to what the human mind can conjure up.

What About All the "Wasted" Time Before Us?

Heeren: How do you answer people who say, "If God created the universe 15 billion years ago, just for us, then think of all the time before us and the billions of empty galaxies around us. Doesn't that seem like an awful waste of time and materials?"

Gange: My answer is, Isaiah 55:8 says, "My thoughts are not your thoughts, and my ways are not your ways. As the heavens are higher than the earth, so are my ways higher than yours and my thoughts than yours." The moment anybody begins to pretend that they can understand the thoughts of God, they have become an imbecile.

And to say, billions of years, so what? Don't you realize that all of that might have happened in a sense that involves warping of spacetime? Time is something very strange—it's a space dimension, actually. . . . If you want a good picture of time, imagine an elastic band, and draw some black marks on the elastic band. Now the interval between any two adjacent marks on an elastic band is the ticktock of a clock. We know that time dilates. We know that we can take a clock on earth and put another clock in one of our satellites, and because it's higher up off the earth—we know that gravity is weak there—we know that the clock runs a little faster. It's known as the equivalence principle in general relativity. . . . Time is so unbelievably not what it appears to be that there really isn't a problem with billions of years. God's outside of it all.

Heeren: Right.

Gange: As you pointed out to me earlier, it's not that space was created with big bang, but space-time. So if you imagine living in flatland, which is one page of the book at a time, then time is the third dimension through the pages of the book. So it's not that one page was created, which is the two dimensions of space confined to this discussion, but the entire book was created. That's the big bang—the entire book "space-time." Well, God's outside of that. What does God care how thick the book is?

We live inside time, and we see things in such a finite, limited
way. But you've got to go by what the data and the evidence
teaches. Remember that Romans 1:20 says that He's going to
hold people accountable, because what can be known about
God—powers, visible deity and so forth—are clearly seen by the
things that are made, so that they're without excuse. That means
that your senses are trustworthy. He's not creating, as some
would have us believe in this puritan age, supernovas that really
didn't explode—you know, this gigantic cosmic conspiracy.

Things to Come

What do these cosmologists have on their minds today? If inventors
create the future in the present, what are these inventors of our latest
theories dreaming up now? Time machines? Solutions to global warming? A 50-cent radio? The post-Hubble-generation of telescopes?
Making contact with extraterrestrials? How about all of the above?

What Will the Next Generation of Telescopes Tell Us?

The Hubble Space Telescope. —*NASA*

How close are we to getting solid evidence to choose between theories of galaxy formation? The repaired Hubble telescope now gives us pictures of galaxies at a redshift of 0.4, meaning that we're looking back about 4.5 billion years, to a time when our own sun was a newborn star. But most models of galaxy formation do not predict that galaxies should look any different at that stage. Astrophysicist James Truran says that a new generation of telescopes is needed to tell us what we need to know.

Truran: What we really need to be able to do is to observe galaxies at redshifts between 1 and 3 or 4, back during the phases where a galaxy was in its early stages of evolution. We would like to know what they looked like at that time—whether they were collections of small clusters and smaller galaxies that were eventually being drawn together and merging into galaxies like our own, or whether they were relatively uniform regions of gas that were collapsing into some relatively more straightforward manner.

Heeren: Does it seem like our technology is going to allow us to find that out in the next couple of years, or are we quite a ways off from that?

Truran: A couple of years—probably not, to be able to look back far enough. HST will give us a lot of information out to a redshift of about 0.5 to 1 in critical regions, but it isn't quite a big enough collector maybe to get us as far back as we would like to go. That may be the next generation of HST-like projects, the next big telescope project for 2005 or 2010 or something.

Putting ET on Hold

Heeren: Is the SETI project—I know it's not funded by the government anymore—but is the Institute making use of Mount Wilson in any way?

Jastrow: Well, we had a SETI grant to pick out stars of about the same age as the sun or older, on the grounds that if they were younger than the sun, they'd have nothing but worms and jellyfish, according to the history of life on this planet. And so we were going to tell NASA which were the best stars to look to. But of course, as you just said, it got canceled.

As I mentioned earlier, Dr. Jastrow went on to say that if life is common in the universe, he fully expects to be hearing from extraterrestrial intelligence soon, one way or another,* because our television and radio signals make us a very conspicuous part of the universe these days. Is anyone out there listening?

---

*SETI continues to be funded privately and similar projects abound.

Who's Warming the Planet?

Jastrow: I'm very interested in my colleague Sallie Baliunas's studies of stars similar to the sun, which are revealing the information that they vary in brightness by an amount that could explain the global warming without recourse to a greenhouse effect.

Heeren: I think I just saw something about that in *Science News*. She must have been talking about that recently.

Jastrow: That's right, there was an item—you keep up on things. So that's my main involvement.

Heeren: And if that turns out to be true, that could actually preclude other causes of global warming that others have been so worried about.

Jastrow: Well, it would answer a puzzling question, because, it would explain why the earth has actually gotten a little bit warmer over the last hundred years or so, without calling on the greenhouse effect. And that would answer the puzzle as to why, if the greenhouse effect is responsible for the earlier warming, in the last decade or two, gases in the atmosphere like CO_2 have been increasing rapidly, and we should see a very pronounced warming at high latitudes, especially in the Arctic and the Antarctic. But satellites have measured temperatures up there and they don't see anything. So the greenhouse effect, for some reason, is missing. And then the question arises, "Okay, what *did* cause the global warming, and Sallie Baliunas's studies provide a tentative answer.

Heeren: That's going to be interesting to see how that develops.

Jastrow: And that could have important implications for government policy in the next couple of decades.

Heeren: Right. Does that have any implications at all for the ozone layer?

Jastrow: She's also independently studying the ozone layer. The two are not directly connected, except they're both alleged to be caused by human factors, human change. And her studies thus far indicate that the kinds of changes that we know to have occurred naturally in the ozone layer—everything we've observed in recent years fits within the envelope of these natural changes. So there doesn't seem to be any evidence, in fact, that human activity is causing any ozone changes.

Heeren: Does that mean that the sun could be having the effect on the ozone changes?

Jastrow: Well, the sun is one of the factors. . . . Ultraviolet radiation from the sun causes the ozone, and you would expect that when the ultraviolet goes up and down, the ozone level would go up and down, and so it does, in fact. It goes up and down every eleven years—and by an amount which is larger than the alleged human cause.

There's something to keep in mind the next time you're sitting in the EPA line waiting for your car to be tested for those deadly hydrocarbons that are allegedly fouling up our planet's ecosystem.

From the Big Bang to the 50-Cent Radio

Heeren: What are you working on now?

Wilson: Oh, not very much in astronomy. There are a number of wireless projects going on here, mostly having to do with inexpensive ways to do wireless things. This is all things for the bottom line at AT&T.

Heeren: Sure. What's your position there at Bell Labs?

Wilson: I'm a department head. And I'm afraid that the most interesting one is one that hasn't been announced yet. But it's a very practical wireless product that we hope will sell in the hundreds of millions. It's a very inexpensive thing. The basic idea is that you can build a two-way radio for fifty cents—if you do it right. . . .*

The astronomy work I've been doing in recent years is associated with interstellar molecules, and observing molecules in interstellar clouds, which are places where star formation's going on. . . . And so we've spent a fair amount of effort in understanding where these clouds are and trying to understand the sequence of events that leads to star formation.

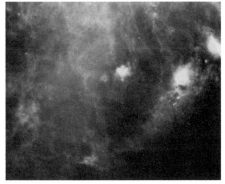

Clusters of young stars and many sites of star formation in regions of heavy interstellar dust near the Pleiades star cluster. —*NASA*

---

*Robert Wilson recently left Bell Labs and went to work for the Smithsonian Astrophysical Observatory. When I last spoke with his office staff (in January 1995), he was at the South Pole doing research that had little to do with the bottom line.

Are Time Machines Actually Possible?

Every time I see car tracks in the snow leading under a garage door, I think of the last scene in *The Time Machine,* when Rod Taylor dragged his invention through the snow and into his laboratory. He wanted to get back to the future (and to Weena, his future girlfriend); and he wanted to be sure he ended up on the right side of a future wall. To my pre-adolescent mind, this was neat stuff.

Later I learned that relativity does actually allow for time travel, of a sort. Time slows down a bit when you approach great masses. Traveling at near light speed could bring you thousands of years into the future in short order. And if you ever managed to fall into a black hole, you might quickly find yourself at the end of the history of the universe (although you'd never make it intact through the event horizon). None of this sounds very practical, if time travel is your goal. Even if you managed to get into the future, you'd be stuck there with no way back. Imagine my astonishment when I asked Alan Guth what he had been working on lately, and he told me that for the past few years he had been researching the possibility of *a true time machine*—one that could bring you back again.

Guth: What I have been working on for the past few years is the presence of what are called closed time-like curves in general relativity. Loosely speaking that translates as time machines. The question is: Can a space-time distort itself so much that it's possible to actually make a complete loop in time, and come back to the same space-time point that you started from?

Heeren: That's real interesting.

Guth: It's been fun.

Heeren: So what kind of conclusions have you come to?

Guth: Well, I guess two conclusions. One is that it's awfully hard to understand under what circumstances these closed time-like curves can and cannot occur. . . . Nonetheless they do arise in situations in classical general relativity, and it's awfully hard to see what prevents them from arising in the real world. But it does seem that there are diverse, complicated, physical effects that prevent them from happening. I was working with three other people. . . . And we were working in particular on a certain class of situations which gives rise to closed time-like curves. And these are time machines if one has things called cosmic strings, which are predicted to exist by a number of particle theories, although nobody knows for sure. . . .

But if cosmic strings do exist, and if you had two infinitely long parallel cosmic strings that passed each other at very high speed, then it turns out that a rocket ship which went around the two strings sufficiently fast could get back to where it took off at the same time that it left—which is what a closed time-like curve means. Now what we discovered was that if one makes certain assumptions about the initial conditions for this kind of a space, then all this could never happen.

So the bottom line is that it could happen in some types of hypothetical universes, but there are other hypothetical universes where it can't happen. So if we're living in one of the hypothetical universes where it can't happen, then we're protected from closed time-like curves.

Disappointed? Perhaps we time-bound creatures are inherently attracted to the idea of time travel because we're very curious about what the future holds for us, or even about the times before and after our brief sojourn on this planet. Or perhaps we enjoy fantasizing about it because we're frustrated with the constraints of having only this threescore-and-ten to do all the things we'd like to do. It almost seems that we were meant for more than just this. As Margaret would say, "Are we living just so we can work, so we can have food and shelter, so we can live, so we can work?" What's the point?

My Online Bible lists 107 references to a word that suggests a solution to the time-crunch problem. The word "everlasting" (or "eternal") suggests what everyone wants more than anything: more time. But even more important, this running theme throughout the Bible is always connected to the one thing that can give us a point to our lives. The word "everlasting" always appears in connection with our Creator, or with those who are close to Him. Who would know the point to living better than the One who made us? And how might we better satisfy our longing to see the future, than by sticking close to Him?

All our explorations into the greatest discoveries of the century have been leading us, not only to some metaphysical concept about a Creator who worked in the remote past, but to the One who desires to make personal contact with us so that He might lead us into the future—and fill our lives with meaning now.

> *You have made known to me the path of life;*
> *you will fill me with joy in your presence,*
> *with eternal pleasures at your right hand.*
> *—Psalm 16:11*

> *Lead me in the way everlasting.*
> *—Psalm 139:24b*

The Bible and the Message from Space

What can we learn from this century's greatest astronomical discoveries?

Discovery 1—1919: During a solar eclipse, Sir Arthur Eddington observed the bending of starlight passing the sun, matching the effect predicted by Einstein's general theory of relativity. If correct, this theory of gravity means that the universe must be expanding. Einstein eventually renounced his belief in an eternal universe and admitted that the universe must have had a beginning. Astrophysicist George Smoot says: "Until the late 1910's, humans were as ignorant of cosmic origins as they had ever been. Those who didn't take Genesis literally had no reason to believe there had been a beginning."

Discovery 2—1927: Astronomer Edwin Hubble discovered that the galaxies are all retreating from us. The more distant galaxies (which show us the more distant past) are retreating from us faster than the nearer galaxies, just as one would expect if the universal expansion is slowing down from an initial surge. Famed astronomer Robert Jastrow says: "The Hubble Law is one of the great discoveries in science: it is one of the main supports of the scientific story of Genesis."

Discovery 3—1965: Arno Penzias and Robert Wilson discovered cosmic background radiation coming from every point in the sky, the remnant predicted by early big bang theorists. Its precise matching with a blackbody spectrum at all frequencies is difficult to reconcile with anything other than a creation event involving the entire universe.

Discovery 4—1970s: Astronomers observe that galaxies are distributed more densely—and quasars become abundant—as they look farther into space, indicating that the universe has changed with time. These observations argue against an eternal cosmos and for a creation event.

Discovery 5—1992: NASA's COBE satellite team discovered the predicted ripples in the cosmic background radiation. George Smoot, the team's leader, called these seeds for future galaxy superclusters "fingerprints from the Maker."

What does all this have to do with the Bible?

Among all the ancient peoples, only the Hebrews got their cosmology right. While the rest of the world believed in a magical, eternal universe that gave birth to the gods, only they believed in an eternal, transcendent God who gave the universe its beginning.

Like every cause, the Cause of the universe must be independent of its effect. Thus, the First Cause must be separate from the universe, not a part of it. From ancient times, the Bible has clearly presented God as non-physical, a Spirit who cannot be contained, even by the heavens. Unlike other ancient religious writings, the Bible prohibited the making of images of God, making it a point to teach that He is not a physical being.

The consensus of modern science is that the universe—and time itself—had a beginning. Nothing that is confined to time could have created the cosmos. God must not only be separate from His creation, but He must exist outside of time. Again, from ancient days, the Bible specifically defined God as the I AM, operating outside of time and existing before the universe He created.

Perhaps the universe had a beginning, but how do we know that it didn't begin by *chance*?

- Stephen Hawking wrote, "If the rate of expansion one second after the big bang had been smaller by even one part in a hundred thousand million million, the universe would have re-collapsed before it ever reached its present state." Slightly *faster* than the critical rate and matter would have dispersed too rapidly to allow stars and galaxies to form. George Smoot describes the creation event as "finely orchestrated."

- Carl Sagan admits: "It is easy to see that only a very restricted range of laws of nature are consistent with galaxies and stars, planets, life and intelligence."

- Hawking cites the critical ratio between the masses of the proton and the electron as one of many fundamental numbers in nature. He adds: "The remarkable fact is that the values of these numbers seem to have been very finely adjusted to make possible the development of life."

- The calculations of Hawking's associate, Roger Penrose, show that the highly ordered (low entropy) initial state for the universe is not something that could have occurred by even the wildest chance.

- When Fred Hoyle calculated the likelihood that carbon would have precisely the required resonance by *chance*, he said that his atheism was greatly shaken, adding: "A common sense interpretation of the facts suggests that a superintellect has monkeyed with physics."

- Princeton physicist Freeman Dyson writes, "The more I examine the universe and the details of its architecture, the more evidence I find that the universe in some sense must have known we were coming." NASA astronomer John O'Keefe says, "It is my view that these circumstances indicate that the universe was created for man to live in."

But isn't religion just a cultural phenomenon, a form of superstition?

For many it is. But perhaps the ultimate superstition is to believe that this physical universe is imbued with mystical powers that enable it to bring itself into existence and then to fine-tune itself.

In the matter of deciding who's running the universe, we all have just three choices: the universe itself, humankind, or God. Because a cause must precede its effect, the first two options violate logic, especially now that we know the universe did not exist in eternity past.

Atheism and pantheism are difficult to reconcile with modern findings. But the Bible fits perfectly, telling us that God is not just a force that's one with the universe, but who is separate from His creation. And like modern physics, the Bible points to a Creator who is super-intelligent, a perfectionist who cares about us a great deal.

Then why would God let our world get into such a mess?

Indeed, the most important implication of a perfectly designed universe is that a perfect Designer would *do* something about the problem of evil in our world.

So what might a super-intelligent, caring Creator do? Make creatures who have no wills of their own, so that they cannot bring evil into His perfect universe? Not if God desired to have an eternal relationship with a people who would willingly return His love. The very idea of a real will to love requires the real possibility of a person's will being used to reject.

So what might be God's options, after his race of free-willed creatures broke the harmony of His universe (as they have obviously done in our case)? He could exterminate them. He could simply overlook their injustices. He could leave them alone to let them try to straighten out their own mess.

But none of these options show the forethought of a perfect, super-intelligent, caring Creator. The Bible, the one book that gave us a true picture of God since ancient times, gives us the one solution that shows great care and forethought, though we might never have thought of it ourselves. What did God do? *He died for us.* He showed both perfect justice and unbounded mercy. And by doing so, He gave those who wanted to be reconciled to Him the chance to be forever changed, to be eventually made into fit company for Him throughout eternity. This was His plan "before time began" (1 Corinthians 2:7).

But what about this biblical idea of God becoming man? What about the concept of sacrifice? Aren't these *primitive* concepts?

If the Creator of the universe wanted to communicate to us (moderns and ancients both) what He is like, how could He show us more clearly than by becoming one of us? If He wanted to communicate to us the seriousness of breaking His moral law, how could He show us more forcefully than by demanding that the most valuable thing in the universe be forfeited as a penalty? And if He wanted to tell us how much He loves us, how could He do so more dramatically than by dying for us?

> *But He was pierced for our transgressions,*
> *He was crushed for our iniquities;*
> *the punishment that brought us peace was upon Him,*
> *and by His wounds we are healed.*
> *We all, like sheep, have gone astray,*
> *each of us has turned to his own way;*
> *and the LORD has laid on Him the iniquity of us all.*
>
> —Isaiah 53:5-6

But giving intellectual assent to the historical idea that Jesus died on a Roman cross won't change anyone's life. Biblical faith always implies personal trust, a personal relationship. This relationship gives us the ability to talk to Him, not just about Him.

This relationship, after all, is the reason He created us. It means our lives aren't pointless; we don't live only to have all memory of us snuffed out in a few generations and throughout eternity. Rather, we find access to eternity through the One who exists outside of time. This is the one relationship that can give our lives lasting value.

Information taken from the book, *Show Me God*. This and other literature available from: Searchlight Publications • 326 S. Wille Avenue • Wheeling, IL 60090 (1-800-743-7700)

INDEX

Fred Heeren has written for both Christian and secular audiences. His work has made use of film, radio, audiotape drama, and theater. Fred's writing credits include a unique mix of published short stories, articles, devotionals, stand-up comedy monologs, and over forty programs for Moody Broadcasting's radio drama series. He serves as president of Searchlight, Inc., a publishing company dedicated to getting into the skeptic's shoes to explore life's big questions. He is best known as founder and president of Day Star Productions, a non-profit Christian ministry incorporated in 1978 to use the media to proclaim the good news of Jesus Christ. His film, *Ordinary Guy*, received the Best Film of the Year Award from the Christian Film Distributors Association.

Fred's research toward his goal of reaching doubters is reflected in the humorous audiotape dramas he has written and produced in recent years: *The Adventures of Leon the Cynic* (dealing with the lives of great scientists who were also outspoken Christians), and *The Skeptic's Guide to the Bible*. The *Wonders That Witness* book/tape and radio series comes out of five years of research and writing to bring the gospel to skeptics. Fred lives with his wife and five children in Wheeling, Illinois.

.